T0248189

Digital Infrastructure and Digital Presence

A Framework for Assessing the Impact on Future
Military Competition and Conflict

JULIA BRACKUP, SARAH HARTING, DANIEL GONZALES

NATIONAL DEFENSE RESEARCH INSTITUTE

For more information on this publication, visit **www.rand.org/t/RRA877-1**.

About RAND

The RAND Corporation is a research organization that develops solutions to public policy challenges to help make communities throughout the world safer and more secure, healthier and more prosperous. RAND is nonprofit, nonpartisan, and committed to the public interest. To learn more about RAND, visit www.rand.org.

Research Integrity

Our mission to help improve policy and decisionmaking through research and analysis is enabled through our core values of quality and objectivity and our unwavering commitment to the highest level of integrity and ethical behavior. To help ensure our research and analysis are rigorous, objective, and nonpartisan, we subject our research publications to a robust and exacting quality-assurance process; avoid both the appearance and reality of financial and other conflicts of interest through staff training, project screening, and a policy of mandatory disclosure; and pursue transparency in our research engagements through our commitment to the open publication of our research findings and recommendations, disclosure of the source of funding of published research, and policies to ensure intellectual independence. For more information, visit www.rand.org/about/research-integrity.

RAND's publications do not necessarily reflect the opinions of its research clients and sponsors.

About This Report

A U.S.-China strategic competition is underway that will be shaped by broad economic, diplomatic, information, and military trends. Networks, cyber, space activities, and the supporting digital infrastructure play a role in this competition given the advantages they can provide. Digital dual-use technologies—often commercially developed and operated—can serve as a force multiplier for military posture and presence and create key intelligence advantages.

This report summarizes a RAND Corporation research project that characterizes the digital infrastructure and digital presence and develops a way to assess the implications of each for near-peer competition and future conflict.

This report may be of interest to practitioners, policymakers, and technologists in the U.S. Department of Defense and Intelligence Community interested in understanding how digital technologies have changed—and will continue to change—competition and conflict between major powers.

The research reported here was completed in October 2021 and underwent security review with the sponsor and the Defense Office of Prepublication and Security Review before public release.

RAND National Security Research Division

This research was sponsored by the U.S. Department of Defense and conducted within the Acquisition and Technology Policy Center of the RAND National Security Research Division (NSRD), which operates the National Defense Research Institute (NDRI), a federally funded research and development center sponsored by the Office of the Secretary of Defense, the Joint Staff, the Unified Combatant Commands, the Navy, the Marine Corps, the defense agencies, and the defense intelligence enterprise.

For more information on the RAND Acquisition and Technology Policy Center, see www.rand.org/nsrd/atp or contact the director (contact information is provided on the webpage).

Acknowledgments

We thank our management team within RAND—Joel Predd, Yun Kang, and Caitlin Lee—for their guidance and encouragement. The analysis contained in this report also benefited from the insights and expertise of several of our research colleagues—Chad Ohlandt, Steven Popper, Michael Kennedy, and Christopher Mouton—throughout this effort. We also thank Scott W. Harold and Marjory S. Blumenthal for their careful and thoughtful reviews of the manuscript. Our report is much improved because of their comments. Finally, we thank Silas Dustin for assisting with the preparation of the draft. The authors alone are responsible for any errors remaining in this report.

Summary

Information and intelligence—and the degree of access to and control of the systems within which the data reside—can yield power and influence at scale. These systems and the networks they create collectively make up what we characterize as *digital infrastructure* (DI). Spawned from internet growth and the interconnectivity of global telecommunication networks, today's DI—and a country's ownership of, access to, and control over it—has emerged as an area of competition between the United States and China. Beijing and Washington rely on DI to support military forces and use its capabilities to expand national power and extend influence globally. Both countries now aim to shape the DI in ways that align with their long-term strategic priorities and interests.

We hypothesize that just as the United States views overseas military presence as a strategic means to compete and deter, so too does the People's Republic of China (PRC) view DI as a strategic means for competition. While the DI affects various dimensions of power, the United States and China approach it differently. China's strategy for achieving long-term strategic goals appears to be heavily—although by no means exclusively—focused on using and exploiting DI; U.S. power continues to stem from a post–World War II model founded on institutions, power projection, and a rules-based liberal order. This divergence in approach creates asymmetries in competitive strategies, means, and mindsets.

In this report, we define DI, characterize the competition for it, and provide evidence showing that how DI evolves has implications for long-term military competition and conflict. We also describe important trends and asymmetries shaping the competition. We conclude with a discussion of implications and opportunities for the U.S. government and the U.S. Department of Defense (DoD).

Purpose of This Report

This emerging competition concerns U.S. national security for several reasons. First, we hypothesize that China's use of DI goes beyond economic and domestic purposes to military and security application on a global scale. Second, given that DoD will become more reliant on DI and that DI will likely affect future warfare significantly, the United States should seek to shape and control it rather than be at the behest of another actor that does do so. In particular, the United States should better understand the emerging DI competition because of its implications for national security and defense.

How this competition for DI influence and control plays out between the United States and China will have implications for future military competition and conflict because of the potential advantages DI offers in key dimensions of power (diplomatic, information, military, and economic). Our analysis yields significant implications and potential opportunities for DoD.

Concepts

In this report, we also identify a linkage between DI, digital presence, and long-term strategic competition. In doing so, we developed and characterized two of these as concepts: DI and digital presence. DI refers to an interconnected set of networks larger than any single country, company, or technology. For the purposes of this analysis, we define *DI* as the network of networks consisting of hardware and software elements that digitally process, store, and transmit data. DI has several components: submarine cables, network points of presence, wireless networks and infrastructure, terrestrial networks and infrastructure, and satellites and terminals. We also identify two elements foundational to DI: microchips and technical standards. Control of DI can lead to what we term *digital presence.*

Digital presence is a nation state's (1) awareness of or visibility into specific activities, actors, and/or information at a time and place without a physical presence and (2) potential to control network resources, networked platforms, and user devices. Digital presence can enable intelligence, economic, technology, and military advantages.

Methodology

The key tenets of our methodology include a basic assessment, a trends and asymmetries analysis, and development of implications and opportunities for DoD. The basic assessment characterizes a competition by identifying important trends, key players, and the overall state of the balance and describing how to think about the competition. The trends analyses highlight interesting and significant strategic and technical asymmetries that shed light on major factors driving and shaping the competition. The trends also begin to outline how the competition evolved and developed over time. Finally, our methodology then uses the analysis from a basic assessment and trends and asymmetries to identify opportunities and considerations for the U.S. military to position itself for future competition. We used this approach to understand and describe an emerging competition, the factors driving it, and potential implications for DoD.

In this report, we begin by defining terms and concepts, describing characteristics associated with the terms and concepts, and describing the assumptions underpinning the assessment. We then present a preliminary assessment of the status for the United States and other key players. We include both structural and technical trends and asymmetries because some set the conditions for technological pursuits, and others are direct outputs of how the technology is advancing. Finally, the discussion of implications and opportunities includes several levels of stakeholders.

To characterize Chinese approaches to DI, we primarily used research from U.S. scholars and subject-matter experts on the People's Liberation Army (PLA) and China more broadly. Any assumptions about either the PLA's view of DI or the Chinese Communist Party's (CCP's) stem from our interpretation of existing secondary sources that leverage PLA writings. We have not, however, consulted or used Chinese writings directly in this report.

Findings and Implications

DI will play a role in shaping the U.S.-China competition. While traditional elements of power (military, economic, information, diplomatic) shape competition and conflict, power centers within each element are shifting toward DI. As a result of these shifts, control of and access to DI offer potential for the United States or China to create competitive advantages—and, in some instances, asymmetric advantages—vis-à-vis each other.

China and the United States recognize DI building blocks and foundational elements (i.e., chips, standards) as important to economic growth; however, the U.S. political-economic model may not be poised to compete with China's approach. China uses a statist economic model that focuses on centrally directed infrastructure investment initiatives. In contrast, after initial government research and development support, the U.S. private-sector developed early iterations of DI with minimal government involvement.

U.S. diplomatic efforts provide structural advantages that may shape the competition for DI in ways favorable to the United States. Longstanding, trusted U.S. partnerships and alliances and the leadership of international institutions provide key structural advantages that may shape the competition in ways favorable to the United States. The United States exerts considerable influence over the international order that China potentially seeks to lead.

How DI evolves may affect warfare in substantial ways. The United States and China have efforts underway to leverage DI and foundational elements for military applications. Because of changes in military systems, operational concepts, and some organizational restructuring tied to DI, we expect DI building blocks and foundational elements to affect the character and conduct of warfare. As a result, ownership of, access to, and control over DI will become increasingly important for military operations.

Ownership of, access to, and control over DI by an untrustworthy actor may introduce risk into a country's national DI. A country or region with a DI dominated by an untrustworthy actor may present unacceptable risks to some military operations or governmental activities. We see the potential for this to occur in areas where Chinese companies have fielded DI.

Characteristics of the future security environment are likely shifting because of DI and have the potential to disrupt the U.S. way of war and traditional approach to power projection. These changes may affect future platforms and capabilities. Key shifts in the future security environment may include the following:

- A country's physical footprint may no longer be sufficient to offer key military advantages in some scenarios
- More than one way to project power exists; previous assumptions associated with power projection are being challenged
- There is a growing rise and importance of dual-use technology and capabilities
- Democratization of signals intelligence and ubiquity of open-source intelligence
- Proliferation and reduction in size of precision strike weapons because of the availability of high-performance chips
- New electronic warfare and cyber capabilities that will render command, control, and communication networks more vulnerable.

DI plays an important but distinct role within the differing U.S. and PRC visions of power projection. The U.S. approach to power projection, predicated on post–World War II assumptions and reliant on traditional military capabilities, differs from China's likely use of traditional and nontraditional means to project power beyond its borders. Furthermore, the roles of DI for the U.S. and PRC visions of power projection also seem to differ. DI and digital presence appear central to PRC power-projection ambitions and potentially challenge U.S. traditional approaches to power projection. China's traditional military power-projection capabilities have grown in recent years but remain limited. However, DI and digital presence may provide the PLA a potential asymmetric means to account for military gaps.

Shaping the long-term strategic competition between the United States and China will require a comprehensive understanding of DI and foundational elements. Military net assessments will benefit from an assessment of DI to account for the role of digital presence in military competition and conflict. Assessments to understand key characteristics of future warfare should also account for digital presence. Characterizing DoD's reliance on and relationship to DI and foundational elements is essential for informing an effective DoD strategy for long-term military competition.

This report serves as a framework for understanding and defining this area of the broader U.S.-PRC competition and lays the foundation for ongoing analysis to understand the implications for military competition and future conflict.

Contents

Figures and Tables

Figures

Tables

CHAPTER ONE

Introduction

Background

Information and intelligence—and the level of ownership of, access to, and control over (OAC) the systems within which the data reside—can yield power and influence at scale. Spawned from internet growth and the interconnectivity of global telecommunications networks, today's digital infrastructure (DI) and a country's OAC of it have emerged as areas of competition for the United States and China. Both countries rely on DI to support military forces and for its capabilities to support national power and extend influence. The two countries aim to shape DI in ways that align with their long-term strategic priorities and interests.

Although to a differing degree, both the United States and China recognize the value of DI for underpinning their long-term strategic goals and position themselves to compete over DI OAC. How this competition for DI influence and control plays out between the United States and China will have implications for future military competition and conflict because of the potential advantages DI offers in key dimensions of power (diplomatic, information, military, and economic), particularly through what we term *digital presence*.

Objective

In this research, we developed a framework to understand how digital technologies have changed competition and conflict between major powers and will continue to do so. For the framework, we developed, characterized, and defined the concepts of *digital presence* and *digital infrastructure*. We intend for this work to serve as the basis for a longer research effort focused on assessing DI and digital presence and the implications for competition and conflict between the United States and China.

Methodology

Our methodology leverages key tenets of traditional U.S. Department of Defense (DoD) net assessments (see Figure 1.1).[1] We leveraged the net assessment approach, given its focus on defining and characterizing key concepts to diagnose issues and implications for DoD and emphasis on understanding the interrelationship between different variables within a competition. In this project, we focused on defining and characterizing DI, digital presence, and what we mean by the emerging *DI competition* between the United States and China.

The key tenets of this framework include a basic assessment, a trends and asymmetries analysis, and development of implications and opportunities for DoD. The basic assessment characterizes a competition by identifying important trends, key players, and the overall state of the balance and describing how to think about the competition. The trends analyses high-

FIGURE 1.1
Key Elements of the Framework

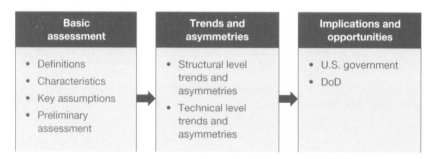

[1] For more information on the history and practice of net assessment, see Mie Augier, "Thinking About War and Peace: Andrew Marshall and the Early Development of the Intellectual Foundations for Net Assessment," *Comparative Strategy*, Vol. 32, No. 1, 2013, and Paul Bracken, "Net Assessment: A Practical Guide," *Parameters*, Vol. 36, No. 1, Spring 2006.

light interesting and significant strategic and technical asymmetries that shed light on major factors driving and shaping the competition. The trends also begin to outline how the competition evolved and developed over time. Finally, the framework then uses the analysis from a basic assessment and trends and asymmetries to identify opportunities and considerations for the U.S. military to position itself for future competition. We used this approach to understand and describe an emerging competition, the factors driving it, and potential implications for DoD.

In this report, we begin by defining terms and concepts, describing characteristics associated with the terms and concepts, and describing the assumptions underpinning the assessment. We then present a preliminary assessment of the status for the United States and other key players. We include both structural and technical trends and asymmetries because some set the conditions for technological pursuits, and others are direct outputs of how the technology is advancing. Finally, the discussion of implications and opportunities includes several levels of stakeholders.

To characterize Chinese approaches to DI, we primarily used research from U.S. scholars and subject-matter experts on the People's Liberation Army (PLA) and China more broadly. Any assumptions about either the PLA's view of DI or the Chinese Communist Party's (CCP's) stem from our interpretation of existing secondary sources that leverage PLA writings. We have not, however, consulted or used Chinese writings directly in this report.

Organization of This Report

In Chapter Two, we define *DI*, the foundational elements, and *digital presence*. In Chapter Three, we characterize the competition for DI and the associated foundational elements. In Chapter Four, we identify and explore trends and asymmetries, both structural and technical. Finally, in Chapter Five, we conclude with a discussion of findings, implications, and emerging opportunities for DoD.

Defining Digital Infrastructure and Digital Presence

In this chapter, we define *DI* and describe the DI building blocks and foundational elements in detail. We also explain the related concept of *digital presence.*

Digital Infrastructure

DI is a network of networks consisting of hardware and software elements that digitally process, store, and transmit data:[1]

- Hardware elements consist of submarine cables, wireline terrestrial network and information infrastructures, wireless network equipment, satellites, and satellite terminals.
- Software elements encompass operating systems, firmware, software, and applications.

Before the first communication satellite (COMSAT) was put into orbit in 1962, networks existed as separate, isolated systems in specific countries or regions. Since then, particularly with the invention of the internet in 1969, the growing interconnection of global telecommunications networks led to DI.

---

[1] This report focuses on the specific hardware and software elements of DI and how various degrees of OAC of DI can confer potential advantages. While the scope does not include an analysis or investigation of the information that gets processed, transmitted, and stored on DI, that information remains a vital part of any research into the effects of DI on competition and conflict.

DI would not exist without two foundational elements: microchips and technical standards. Microchips are essential components for all parts of DI. Technical standards, which are also essential, provide network design and function information that enables networks to interconnect, operate securely, and provide a range of services. An actor with control over DI microchips or technical standards could change the function and security properties of DI components.

DI consists of four main building blocks:

- **terrestrial**: wireline and fiber-optic networks and the information infrastructure these networks link, including data centers, cloud computing infrastructure, personal computers (PCs), and other user devices
- **cellular:** wireless network infrastructure, including mobile devices
- **space:** satellites and satellite terminals
- **international:** submarine cables and telecommunications network points of presence (PoPs).

Each building block has extensive supply chains and numerous parts and pieces, each of which are designed and manufactured by companies. As we discuss in this section, varying types of OAC of these building blocks offer opportunities for a digital presence. Furthermore, OAC of DI and foundational elements provides the ability to control the "generation, transmission, access to, exploitation, interpretation, and credibility of information."[2] We also refer to the degree of OAC—largely determined by a country's market presence in each building block—as a country's DI footprint.

Terrestrial

The terrestrial building block consists of wireline (e.g., traditional twisted pair wire, coaxial cables and fiber optics) networks and the end points or information infrastructure that these networks link, including data centers, individual PCs and other types of end user devices. Data centers include

---

[2] Brad D. Williams, "FCC's Carr: Close Chinese Backdoors into U.S. Networks," Breaking Defense website, April 1, 2021.

private, public, and hybrid cloud computing centers. Terrestrial telecommunications have evolved as new technologies emerge, leading to dramatic changes in terrestrial network infrastructure from copper wire public switched telephone networks (providing local and long-distance voice calls) to the modern internet (providing a wide range of applications and services to private corporations; consumers; and critical infrastructure, such as water utilities and the electric power grid).

Ownership rules and patterns for telecom companies differ from country to country. Foreign ownership of telecoms remains limited in many but not all Western countries. For instance, U.S. telecom companies, such as AT&T and Verizon, appear to own few overseas telecom assets.[3] However, within the European Union, companies with a headquarters in one country can own telecommunications in others. Other European companies, such as Vodafone (United Kingdom [UK]), have owned telecom operators in Germany, India, and Egypt.[4]

How different countries use terrestrial infrastructure also varies. Many governments can access private messages only for law enforcement purposes. For example, in the United States, a law enforcement agency must have a wiretap order, per the Communications Act of 1934, to access information and communication technology networks.[5] In China, however, all firms, including telecoms firms, must cooperate with the government, per the People's Republic of China (PRC) 2017 National Security Law (whether for overseas or domestic matters).[6]

Furthermore, the type of telecom infrastructure differs, as do the companies that own and operate it. U.S. companies, such as Cisco and Juniper,

---

[3] The Federal Communications Commission (FCC) requires U.S. international carriers to report any affiliation with a foreign carrier (FCC, "Foreign Ownership Rules and Policies," webpage, undated).

[4] Vodafone, a UK telecom, provides telecom services in 49 countries with partner companies (co-ownership) and provide internet connectivity in 74 countries (Vodafone, "Where We Operate," webpage, undated).

[5] The Communications Act of 1934, as codified in 47 U.S.C. § 151 et seq.

[6] James Andrew Lewis, "5G: The Impact on National Security, Intellectual Property, and Competition," statement before the Senate Committee on the Judiciary, Washington, D.C., May 14, 2019.

supply routers and switches to telecom operators but not all the equipment used in telecommunications networks in the global market. On the other hand, Chinese companies (Huawei and Zhongxing Telecommunications Equipment Corporation [ZTE]) now supply complete sets of telecom infrastructure in many countries. We refer to a complete set of terrestrial telecom wireline network technologies as consisting of

- routers and switches that transmit voice and data traffic
- twisted-pair copper wire links (predominantly in older telecommunications networks)
- co-axial cable links
- fiber-optic links for the high-speed backbone of telecommunications networks.

The terrestrial network also includes such key connected endpoints as network control centers, data centers, and cloud computing centers; PCs; and other computing and communications devices. These network endpoints store data and are where users create data using applications. For these reasons, we call this part of the terrestrial network the *application layer*.[7]

Analysis of Chinese telecom supply relationships overseas indicates that the PRC may use terrestrial telecom infrastructure to maintain situational awareness of strategic sea lines of communication.[8]

Cellular

The cellular building block consists of the wireless network infrastructure, which includes mobile devices. Key components of such networks include

- wireless base stations that transmit and receive signals from mobile phones and other devices

---

[7] Andrew S. Tanenbaum, *Computer Networks*, 2nd ed., Hoboken, N.J.: Prentice Hall, 1989.

[8] Executive Research Associates Ltd., "China's Telecommunications Footprint in Africa," in Executive Research Associates Ltd., *China in Africa: A Strategic Overview*, Craighall, South Africa, October 2009.

- core network servers that manage network resources and account billing
- backhaul networks that link base stations to the core wireless network and to the global telecommunications network.

Wireless network ownership and supplier patterns largely mirror those of telecommunications.[9] The United States permits foreign companies to partly or wholly own wireless carriers operating in the United States. For example, Deutsche Telekom, a German company, has an ownership stake in T-Mobile, which operates throughout the United States. In contrast, no wireless carriers in China have foreign ownership. U.S. firms have few, if any, foreign wireless carriers.

However, network monitoring and data-collection capabilities and rules differ from those that govern terrestrial telecommunications. Cellular communications have less clearly defined distinctions between legal and illegal monitoring. Unlike the more-stringent rules that often exist in terrestrial telecommunications (e.g., the United States requires a wiretap order), an actor can eavesdrop on third-generation and fourth-generation (4G) wireless calls without direct access to network infrastructure (the third-generation Short Message Service [SMS], for instance, is not encrypted).[10] Fifth-generation (5G) wireless networks will be more secure by design; however, the infrastructure supplier of the 5G network will likely maintain privileged access to network control systems.

China may develop a dominant supplier role globally in 5G networks, potentially enabling widespread surveillance and monitoring of wireless network traffic and control of wireless networks in crisis or conflict, even via third-party providers. Huawei, for example, is a supplier to Deutsche Telekom and may have privileged network access to its systems. Privileged network access could enable Huawei to exploit a trusted relationship between Deutsche Telekom and T-Mobile networks. China could exploit

---

[9] The FCC requires U.S.-international carriers to report any affiliation with a foreign carrier (FCC, undated).

[10] Chinese firms supply 70 percent of 4G infrastructure in Africa (Amy Mackinnon, "For Africa, Chinese-Built Internet Is Better Than No Internet at All," *Foreign Policy*, March 19, 2019).

such a trusted supplier role to monitor customer communications in the competition phase or to disable network services to specific customers in conflict. Recent U.S. policy moves attempt to prevent Huawei from taking a dominant supplier role in the United States and allied countries.[11]

Space

The space building block consists of satellites and satellite terminals. Commercial COMSATs, which are privately owned satellites, provide network communications independent of terrestrial and wireless networks. Four very large providers provide 50 percent of COMSAT capacity: Intelsat, Telesat, Eutelsat, and SES S.A.[12] Currently, China is not a major player in the international COMSAT market, as either a supplier or a constellation owner. For example, China leases Western COMSAT capacity in the South China Sea.[13] On the other hand, the major suppliers of COMSATs and COMSAT components include U.S. aerospace companies and European satellite suppliers.

Because of their long-range and wireless capabilities, COMSATs are especially useful for ground mobile, maritime, and aircraft communications and are used for both military and commercial sectors. Satellite networks can be entirely independent of terrestrial networks, submarine cables, and wireless networks, which potentially provides greater resilience. Such resilience may be an important capability for military forces.

Satellite technology has never been fully commercial and has always had some dual-use technology elements. The International Traffic in Arms Regulations (ITAR), a policy designed to prevent the transfer of critical technologies, has worked relatively well in preventing the transfer of U.S. and European satellite technology to China. As of this writing, the United States has a large share of the COMSAT manufacturing market relative to China's. In addition, as of 2022 China lacks a COMSAT constellation in low earth

[11] Karen Freifeld, "Biden Administration Adds New Limits on Huawei's Suppliers," Reuters, March 11, 2021.

[12] "Commercial Constellations Don't Live Up to the Hype: Euroconsult," *Satellite Pro Middle East*, December 13, 2020.

[13] Brian Spegele and Kate O'Keeffe, "China Exploits Fleet of U.S. Satellites to Strengthen Police and Military Power," *Wall Street Journal*, April 23, 2019.

orbit but has announced plans to develop one; U.S. efforts in this area have continued to grow through such companies as SpaceX, which now has over 1,000 satellites in low earth orbit.[14]

International

The international building blocks consist of (1) submarine cables and (2) telecom carrier network PoPs.

Submarine Cables

Submarine cables provide high-speed, high-capacity telecom links, typically between national networks of countries on different continents or between national digital infrastructures (NDIs) of different countries. Submarine cables have significantly more bandwidth capacity than satellites and carry 95 to 99 percent of internet traffic between countries.[15] Cables also transport data for international air traffic control, electronic banking and debit card use, phone lines, email, and many other services. Current capacity needs (because of video streaming demands, for example) have created a growing demand for more submarine cables.

Given the potential implications of cable ownership, we conducted an initial survey of submarine cables in the Pacific. We used TeleGeography's map of global submarine cables to survey the cables that connect North America and Asia.[16] This analysis showed that U.S. and Chinese companies own the majority of submarine cables in the Pacific, with Chinese companies having increased ownership in recent years, so we focused on cables with landing sites in the United States and China.

Cable landing sites offer key collection and transmission points where an actor could monitor; copy; and, in some countries, censor international

---

[14] "Top 10 Satellite Manufacturers in the Global Space Industry 2018," *Technavio Blog*, October 9, 2018.

[15] Doug Brake, "Submarine Cables: Critical Infrastructure for Global Communications," Washington, D.C.: Information Technology & Innovation Foundation, April 2019.

[16] TeleGeography, "Submarine Cable Map," webpage, last updated November 15, 2021.

traffic.[17] Cable landing sites also provide convenient locations for national firewalls of some nations, such as China, to prevent access to forbidden websites, press reports, etc. Of note, recently laid cables landing in China and the United States have Chinese ownership. Table 2.1 shows major submarine cables in the Pacific—current and planned—including information about their names, owners, landing sites, and design capacity. The table shows that Chinese owners tend to be Chinese telecom companies, such as China Telecom, that own terrestrial or wireless DI, while U.S. owners tend to be application-layer companies, a difference we discuss further in Chapter Four. The red and blue boxes denote owners and landing sites, with China shown in red and the United States shown in blue.

The table illustrates that most cables have dual Chinese-U.S. company ownership. One cable, SXS, has sole U.S. company ownership and connects California and Guam. Another cable of note, Asia Direct Cable (ADC), has U.S. and Chinese company ownership but has no U.S.-based landing sites. We infer from this initial survey of Pacific cables that U.S. and Chinese companies remain interested in owning submarine cables, even when the cables do not have a landing site in a company's home country.

Control (often conferred through ownership) of cables and landing sites can provide system owners with opportunities for covert monitoring, exfiltration, and disruption. For example, a data stream can be diverted, copied, and/or subjected to traffic analysis.[18] Cable ownership or landing site control can also enable unsupervised access to cable data streams.

Submarine cables have been targeted in the past and will likely remain an attractive target in the future.[19] Exploitation, disruption, or destruction

[17] Brake, 2019, p. 6; Helene Fouquet, "China's 7,500-Mile Undersea Cable to Europe Fuels Internet Feud," Bloomberg, March 5, 2021.

[18] However, message encryption provides some protection against compromise of message contents.

[19] For past examples of cables being targeted, see Sebastien Roblin, "Russian Submarines Could Be Tampering with Undersea Cables That Make the Internet Work," *National Interest*, September 27, 2020; John Mooney, ""Russian Submarines 'Target Subsea Cables' off Coast of Kerry," *The Times* (London), August 16, 2020; Jonathan Reed Winkler, "Silencing the Enemy: Cable-Cutting in the Spanish-American War," War on the Rocks website, November 6, 2015; Gordon Corera, "How Britain Pioneered Cable-Cutting in World War One," BBC News, December 15, 2017; Gaynor

of submarine cables can enable or reduce the probability of successful intelligence operations and could degrade military command and control capabilities in conflict.

Network Points of Presence

Firms that own a network typically control and monitor their own telecommunications networks. The firms employ a combination of firewalls, access controls, and authentication mechanisms to ensure that only authorized users can change the configuration of the network; monitor the network; and, if necessary, copy or divert network traffic (e.g., to enable wiretap systems in response to a legal wiretap order).

Network PoPs connect independently controlled autonomous networks and facilitate the transfer of network traffic between the networks. PoPs consist of hardware and software elements.

A network PoP of a foreign telecom enables that telecom carrier to

- serve as a relay point or intermediary between one or more domestic telecom carriers
- provide a secure relay point to cyberattackers who need to transmit stolen information to their parent organizations
- provide a security boundary for interconnection points between independent networks that have independent network control and monitoring capabilities
- establish a trust relationship with other international telecom companies per standard telecom protocols, i.e., using the Border Gateway Protocol (BGP).

China and the United States have approached PoPs differently, with China focusing more on developing international PoPs that are embedded in foreign networks. China has encouraged its "big three telecom carriers

---

Dumat-ol Daleno, "CNMI Declares Emergency," *Pacific Daily News*, July 16, 2015; Jonathan Barrett and Yew Lun Tian, "Pacific Undersea Cable Project Sinks After U.S. Warns Against Chinese Bid," Reuters, June 17, 2021.

TABLE 2.1

Key Current and Planned Submarine Cables in the Pacific

Cable Name	Owners	Landing Sites	Start Date	End Date	Design Capacity (Tbps)
Trans-Pacific Express	■■ China Telecom, China Unicom, Chunghwa Telecom, KT, Verizon, NTT, AT&T	■■ Taiwan; USA (California); Japan, South Korea; China	2008	2033	25.6
Pacific Light	■■ Google, Facebook, Pacific Light Data Communications, Dr. Peng Telecom & Media Group Co.	■■ Philippines; Taiwan; Hong Kong; USA (California)	2020	2044	144.0
Southeast Asia-Japan Cable 2	■■ China Mobile, Chuan Wei, Chunghwa Telecom Co., Facebook, KDDI, Singapore Telecommunications Limited, SK Broadband, VNPT	■■ Taiwan; Cambodia; Vietnam; China; Hong Kong; Singapore; Thailand; Japan; South Korea	2020	2045	144.0
Hong Kong–America	■■ Facebook, Telstra, TATA, China Telecom, China Unicom, RTI	■■ USA (California); Hong Kong	2021	2046	76.8
Jupiter	■■ Amazon Web Services, Facebook, NTT, PLDT, PCCW, Softbank Corp.	■■ USA (Tierra Del Mar; Hermosa Beach); Philippines; Japan	2021	2046	60.0
SXS	■ RTI Connectivity	■ USA (California); Guam	2022	2047	96.0
ADC	■ ADC Consortium, CAT Telecom, China Telecom Corporation, China Unicom, PLDT, Singtel, SoftBank, TATA Communications, Viettel	■ Singapore; China; Japan; Philippines; Thailand; Vietnam	2022	2047	140.0
CAP-1	■■ China Mobile, Facebook, Amazon Web Services	■■ USA (California); Philippines	2022	TBD	TBD

SOURCE: TeleGeography, 2021. Data as of August 2021.

NOTES: ■ = Chinese owner or landing site in China; ■ = U.S. owner or landing site in United States. KT = KT Corp.; NTT = Nippon Telegraph & Telephone Corp.; KDDI = KDDI Corporation (KDDI Kabushiki Gaisha [株式会社]; SK = SK Telecom Co., Ltd; VNPT = Vietnam Posts and Telecommunications Group; RTI = RTI Group Companies; PLDT = PLDT Inc.; PCCW = PCCW Limited; CAT = CAT Telecom Public Company Limited; TATA = Tata Communications Limited.

to expand overseas."[20] China Telecom, for example, has established more than 200 PoPs outside China, including at least three PoPs in the United States. Because PoPs have dual-use potential, the presence of foreign PoPs in another country could pose national security concerns.

BGP enables malicious actors to "target, alter, block, and re-route" communications.[21] BGP can enable a cyberattacker to implant malware in legitimate messages sent to defense contractors or government agencies. Chinese telecommunications have reportedly employed BGP to hijack U.S. telecommunications and route them to China.[22] Such an approach aligns with Chinese efforts to engage in a "systematic campaign to steal U.S. advanced weapons technology from U.S. defense contractors."[23] Furthermore, in a conflict, China could direct its state-owned network PoPs overseas to disrupt or block network traffic in host countries or to implant malware in telecommunications networks or other critical infrastructure.

Digital Presence

Access to the DI building blocks we have defined and discussed offers a form of digital presence that can be used for military and intelligence advantage. We define *digital presence* as a nation-state's (1) awareness of or visibility into specific activities, actors, and/or information at a time and place without a physical presence and (2) potential to control network resources, networked platforms, and user devices. Countries strive to control and maintain awareness of and access to their NDI and data and to limit unauthorized control and data access by adversaries. Access to or control of DI building blocks can enable a country to have a digital presence in another

---

[20] U.S. Senate, Permanent Subcommittee on Investigations, *Threats to U.S. Networks: Oversight of Chinese Government-Owned Carriers*, staff report, Washington, D.C., June 9, 2020.

[21] Executive Branch Recommendations re China Mobile USA (redacted), referenced on pp. 29–31 of U.S. Senate, Permanent Subcommittee on Investigations, 2020.

[22] U.S. Senate, Permanent Subcommittee on Investigations, 2020.

[23] According to a recent Senate report, China has reportedly stolen the plans for the F-22, F-35, and C-17 (U.S. Senate, Permanent Subcommittee on Investigations, 2020).

country's NDI. Countries can have a digital presence without a significant DI footprint—the level of OAC of DI building blocks—but it is limited and episodic (e.g., cyber intrusions and attacks). However, a significant DI footprint may provide digital presence at a much greater scale. Digital presence can enable intelligence, economic, technology, and/or military advantages.

Digital presence can also act as a force multiplier for militaries. Traditionally, a country's military footprint—its posture, presence, and capabilities— is an indicator of the global military balance of power. A country's military strength, ability to project power and influence through military capabilities, overseas presence, and partnerships all feed into its national strength and ability to deter threats, reassure allies and partners, and build confidence and support at home. Increasingly, a country's DI footprint provides an additional indicator of power and influence because that footprint can provide a digital presence in geographic areas in the absence of, or in addition to, a physical presence that has the potential to serve as a force multiplier.

As mentioned earlier, foundational elements—microchips, standards— are present in all DI building blocks. Microchips and standards include intellectual property, software, and their associated supply chains. Similar to DI building blocks, OAC of foundational elements offers another avenue for digital presence. A country's OAC of foundational elements could be reflected in several ways:

- **influence over a standard:** the ability to ensure the integrity of or to subvert a standard[24]
- **ownership of intellectual property:** the ability to influence or manipulate the design of a technical subsystem or component

---

[24] International technical standards for networks and information technologies are often produced in working groups using consensus-based or voting methods. For these reasons, it can be difficult for an actor to obtain control over or to influence the content of a standard. On the other hand, technical standards are often based on patents, which do not have to reveal details of the technologies, software, or algorithms used to achieve a specific technical capability specified in the technical standard. In this case, a system that complies with the technical standard may have to incorporate a component, typically a microchip, built or designed by the patent holder. An untrustworthy patent holder could insert malicious hardware or software into the component and, thereby, influence or subvert systems that comply with the technical standard.

- **ownership of software and chip design:** the ability to control access to intellectual property, software source code, system design, or design tools and skills
- **ownership of a state-of-the-art foundry:** the ability to ensure the integrity of a chip or to subvert the chip design
- **ownership and/or control of applications:** the ability to control access to device sensors and activity states; device owner data (personally identifiable information); and/or other information the device owner obtains, including sensitive government or private corporate information.

As noted earlier, OAC of DI ties directly to opportunities for obtaining and maintaining digital presence. However, the type and degree of digital presence an actor has depends on several factors or activities. We show these primarily in column one of Table 2.2, which details the type of operational activities associated with DI building blocks, what degree of presence the activity affords, and what capabilities it provides to the actor.

An information technology (IT) or network owner typically has full control of and access to a network (unless functions are outsourced), which provides the ability to control and monitor IT and network infrastructure. A prime contractor that supplies full IT or network systems has a varied degree of access, which could enable overt control and access if functions are outsourced to the prime or covert control and access if the prime configures products to allow remote control or access. A manufacturer of phones, PCs, and/or servers has initial control of and access to user devices, which may allow control and access to its products after they are sold to users, and to products that use the supplier's software. A service provider that operates or supports operations may have full control and access if the owner does not monitor operations; in this case, the lack of monitoring by the owner enables access and possible remote control of network IT, the telecom network, and/or end user devices.

Table 2.2. includes external agents, such as cyberattackers, who exploit weaknesses or vulnerabilities in a corporate or government network or a user device to conduct espionage or intellectual property theft or to attempt to blackmail an organization for ransom (e.g., a ransomware attack). In this case, the primary target may not be the DI or DI owner, but the attack will

TABLE 2.2
Methods to Achieve Digital Presence in Digital Infrastructure Building Blocks

Function	Definition	Degree of Presence	Control and Access Capabilities
Owner	IT or network owner	• Typically, full control and access but can be limited	• Controls and monitors IT and network infrastructure, unless some or all control or monitoring functions are outsourced
Network supplier—prime contractor	Supplies complete IT or network systems	• Full, partial, or no control • Full, partial, or no access	• Overt control and access if functions are outsourced to prime supplier • Covert control and access if prime configures its products for remote control or access (front and back door access)
Supplier—end user devices	Manufacturer of phones, PCs, servers	• Control limited to user devices • Access limited to user devices	• Control and access limited to supplier products or products that use supplier software (e.g., operating systems or apps)
Service provider	Operates or supports operation	• Possible full control and access if owner does not monitor operations	• Enables front door access and possible remote control of IT or network
Cyberattackers	External agents who conduct espionage or intellectual property theft	• Full or partial control and access of corporate or government networks or user devices but usually only for short periods	• Capability to find and exploit cyber vulnerabilities in corporate or government networks or user devices • Ability to exploit the DI to support the cyberattack and to exfiltrate data from the target network

Table 2.2—Continued

Function	Definition	Degree of Presence	Control and Access Capabilities
Installer	Deploys IT or network	• Limited—one-time access for hardware, unless implant is installed	• Enables modification of infrastructure during installation • May enable covert access or sabotage of infrastructure
Supplier—subcontractor	Supplies IT or network components	• Limited control • Limited access	• Supplier may remotely control and access its own products or use its products to expand control to larger parts of the network
Microchip designer	Designs microchips for DI	• No or limited control or access	• Could implant malicious circuitry or functionality into microchips used in DI applications
Technical standard holder	Influence over a technical standard	• Limited control or access	• If the standard is implemented using only the standard holder devices, the holder could implant malicious circuitry or functionality into DI components or applications
Microchip fabricator	Makes microchips for DI	• No or limited control or access	• Could implant malicious circuitry or functionality into microchips used in DI applications

NOTE: Shaded rows = DI supply chain. Control or access enables monitoring, copying, or change of data. Control enables system or device shutdown.

inevitably use the DI to facilitate the attack. In this case, the cyberattacker may not obtain a digital presence in the DI but will achieve a temporary digital presence in the target network or user device. Once the attack is discovered and the compromised information system remediated, the attacker loses its digital presence at the target.

In terms of DI supply chain, an installer that deploys IT on the network has limited, one-time access to the hardware, which enables the modification of infrastructure during installation, which in turn may also enable long-term covert access, sabotage, or control of infrastructure if the installer inserts malware or an implant into the system. Subcontractors that supply IT or network components have limited control and access—they may remotely control and access their own products but could use their products to expand control to larger parts of the network, through the use of malware or implants. A microchip designer has limited or no control of the target system, but it could enable later access to the system—through the use of malicious circuitry implanted into microchips, which are used in the DI. Finally, a microchip manufacturer also has limited or no control or access— although it, too, could potentially implant malicious circuitry into the chip during the manufacturing process.

CHAPTER THREE

Characterization of the Competition

In this chapter, we characterize the ongoing competition for DI by identifying key assumptions underpinning the competition and contextualizing it within the broader strategic competition between the United States and China. Several key assumptions underpin the U.S.-China competition for DI:

- China is pursuing a military power-projection capability.
- The U.S. military has a growing reliance on DI.
- A country's degree of OAC of DI building blocks and foundational elements can yield tremendous benefits or risks in military competition and conflict.

China's Power-Projection Ambitions

PLA modernization efforts, publications, and broader CCP national security goals demonstrate China's intent to develop a global power-projection capability.[1] Despite China's interest in pursuing military power projection, PLA traditional power-projection capabilities still lag relative to the United States.[2] We hypothesized that China may be pursuing asymmetric or non-

---

[1] Chad Sbragia, written testimony, in U.S.-China Economic and Security Review Commission, "China's Military Power Projection and U.S. National Interests," hearing, Washington, D.C., February 20, 2020, pp. 2–3.

[2] Patrick M. Cronin, Mira Rapp-Hooper, Harry Kresja, Alex Sullivan, and Rush Doshi, *Beyond the San Hai: The Challenge of China's Blue-Water Navy*, Washington, D.C.: Center for a New American Security, 2017; Phillip C. Saunders, *Beyond Borders: PLA Command and Control of Overseas Operations*, Washington, D.C.: National Defense University,

traditional means to close gaps in global power-projection capabilities and better address the threat U.S. military forces pose. China's pursuit of dual-use global investment in physical infrastructure and DI to support advancements in information and intelligence operations provides evidence in support of this hypothesis. To unpack this assumption, we next briefly review the core tenets of traditional military power projection.

Traditional Military Power Projection

Academics and analysts describe traditional military power projection as a country's use of military posture, presence, and capabilities to deploy military force beyond its borders and territorial waters. Power-projection capabilities typically include those with tremendous reach (i.e., the ability to transit long distances) at speed, such as long-range strike assets, strategic airpower assets, and a blue-water navy. Overseas bases and military installations further support such capabilities as large-scale expeditionary deployments to project power beyond a state's borders and require overseas basing and access. Posture, force presence, and the capabilities that underpin force employment align to support strategic goals and national security threats. Developing, maintaining, and employing a power-projection capability require immense resources and partnerships with other countries to enable access for military forces and capabilities. Conversely, such an overseas presence reassures partners and allies with shared threats and interests.

July 2020; Mark R. Cozad and Nathan Beauchamp-Mustafaga, *People's Liberation Army Air Force Operations Over Water: Maintaining Relevance in China's Changing Security Environment*, Santa Monica, Calif.: RAND Corporation, RR-2057-AF, 2017; Cristina L. Garafola and Timothy R. Heath, *The Chinese Air Force's First Steps Toward Becoming an Expeditionary Air Force*, Santa Monica, Calif.: RAND Corporation, RR-2056-AF, 2017. We define *military power projection* as the "deployment of military force beyond a state's borders or territorial waters" and *military power-projection capabilities* as "the force structure required to deploy military force over distance" (Jonathan N. Markowitz and Christopher J. Farriss, "Power, Proximity, and Democracy: Geopolitical Competition in the International System," *Journal of Peace Research*, Vol. 55, No. 1, 2018, p. 79).

China's Power Projection

China is pursuing traditional aspects of military power projection (e.g., aircraft carriers and overseas bases) in addition to nontraditional, or potentially asymmetric, ways to project power and influence.[3] The PLA is developing a blue-water navy, expeditionary air capabilities, and overseas presence but currently lacks robust logistics networks and key enablers (e.g., strategic airlift and tanker refueling).[4] In 2017, China established its first overseas military base in Djibouti; in 2019, it was reported that the PLA arranged access to Cambodia's naval base, Ream, in the Gulf of Thailand.[5] Furthermore, the PLA Navy (PLAN) has begun resourcing and developing expeditionary and logistics platforms required for a blue-water navy. Some estimate that, by 2035, the PLAN will have the following:[6]

- up to 25 Type 052D and ten to 12 Type 055 destroyers
- two to four more aircraft carriers
- six to eight Type 075 landing helicopter docks
- 28 Type 054 frigates
- eight to ten Type 071 amphibious assault ships
- 15 Type 072A amphibious warfare ships.

---

[3] Although China has established only one military base to date, many scholars note the likelihood that China is exploring others. For more on this, see Chad Peltier, Tate Nurkin, and Sean O'Connor, *China's Logistics Capabilities for Expeditionary Operations*, Jane's for the U.S.–China Economic and Security Review Commission, 2020; Christopher D. Yung, Ross Rustici, Scott Devary, and Jenny Lin, *"Not an Idea We Have to Shun": Chinese Overseas Basing Requirements in the 21st Century*, Washington, D.C.: National Defense University, 2014; Peter A. Dutton, Isaac B. Kardon, and Connor M. Kennedy, *Djibouti: China's First Overseas Strategic Strongpoint*, Newport, R.I.: U.S. Naval War College, April 2020.

[4] Gregory B. Poling, written testimony, in U.S.-China Economic and Security Review Commission, 2020, p. 2; Chad Peltier, written testimony, in U.S.-China Economic and Security Review Commission, 2020, p. 4; Kevin McCauley, written testimony, in U.S.-China Economic and Security Review Commission, 2020.

[5] Poling, 2020, pp. 5–6; Jeremy Page, Gordon Lubold, and Rob Taylor, "Deal for Naval Outpost in Cambodia Furthers China's Quest for Military Network," *Wall Street Journal*, July 22, 2019.

[6] Peltier, 2020, p. 3.

These planned platforms demonstrate China's intent to develop expeditionary power-projection capabilities that will enhance the PLA's ability to deploy military force beyond its borders and territorial waters.

Despite these efforts, key gaps in the number of platforms, strategic lift, tanker, and logistics capabilities limit the PLA's ability to sustain "a protracted overseas campaign."[7] Additionally, the PLA lacks a mature carrier strike capability and has a limited number of bombers and fifth-generation fighters that could operate effectively in a contested environment.[8] These remaining capability gaps in power-projection platforms suggest that the PLA will not achieve a global power-projection capability even by 2035.

However, when looking at how China has developed, resourced, and leveraged various investment and infrastructure efforts globally, a more nuanced power-projection story emerges. Table 3.1 shows that, although continuing to indicate a keen interest in traditional military power projection, the PLA has also pursued more nontraditional power-projection initiatives, as shown in the last three columns.

Table 3.1 outlines China's investment in civilian infrastructure, gaining a global presence through Belt and Road Initiative (BRI) investments and commercial partnerships. We hypothesized that China's dual use of investment and infrastructure—both physical and digital—serves as an enabler to project power in nontraditional ways and supplement key military gaps. China overtly discusses the intent, if needed, to use civilian infrastructure for military applications.[9]

In some instances, the commercial value of certain PRC infrastructure investments appears murky, indicating potential secondary motives for investment (e.g., military use). Some scholars question the commercial value of the Indo-Pacific BRI infrastructure, arguing that the PRC likely intends to use it for military purposes. For example, China funded the development of the Gwadar port in Pakistan, which could have commercial *and* military value because of its strategic location as a potential logistics port for the

---

[7] Peltier, 2020, p. 4.

[8] Peltier, 2020, pp. 4–5.

[9] Peltier, 2020, p. 8.

TABLE 3.1
China's Potential Power-Projection Efforts

	Military Presence	Military Platforms	Global Infrastructure (dual-use)	Global Investment (dual-use)	Global DI (dual-use)
Evidence	• Overseas basing and regional installations • Location of military capabilities	• Aircraft carriers • Stealth aircraft • Advanced cruise missiles • Ballistic and hypersonic weapons	• Ports • Airports • Railroads • Highways and roads • Government buildings	• Latin America • Africa • Europe • Mideast	• 5G networks • Cellular phones and infrastructure • Telecommunications • Submarine cables • Cybersecurity products • Big data and artificial intelligence (AI) • Satellites
Trend	• Pursuit of a blue-water navy, a more-strategic air force, and overseas basing and military installations	• Increased emphasis on power-projection capabilities, but major gaps remain	• Large global infrastructure efforts that have dual-use possibilities	• Increased foreign economic investment that could be leveraged for military use	• Increased market share of PRC companies in key digital technologies • Space-based intelligence, surveillance, and reconnaissance and communications • Digital technologies could potentially offset gaps in military capabilities

SOURCES: Office of the Secretary of Defense, *Military and Security Developments Involving the People's Republic of China 2020: Annual Report to Congress*, Washington, D.C.: U.S. Department of Defense, 2020; Timothy R. Heath, *China's Pursuit of Overseas Security*, Santa Monica, Calif.: RAND Corporation, RR-2271-OSD, 2018; Dennis C. Blair, written testimony, in U.S.-China Economic and Security Review Commission, 2020; Kristen Gunness, written testimony, in U.S.-China Economic and Security Review Commission, 2020; McCauley, 2020; Paul Nantulya, written testimony, in U.S.-China Economic and Security Review Commission, 2020; Peltier, 2020; Sbragia, 2020; Poling, 2020; Cynthia Watson, written testimony, in U.S.-China Economic and Security Review Commission, 2020; Organisation for Economic Co-operation and Development, *China's Belt and Road Initiative in the Global Trade, Investment and Finance Landscape*, Paris, 2018; World Bank Group, *Belt and Road Economics: Opportunities and Risks of Transport Corridors*, Washington, D.C., 2019.

PLAN.[10] Additionally, China's civilian international airport project in Dara Sakor, Cambodia, has infrastructure that could support military use. Satellite imagery indicates that the aircraft turn bays may be too small for commercial aircraft and better suited for fighter aircraft. The airport also has a 3,400-m airstrip.[11] Although these accounts remain speculative, we do see explicit evidence of China using civilian infrastructure for dual-use (military and security, commercial) and discuss these below.

China's activities in Africa over the past two decades illustrate a history of leveraging civilian infrastructure for dual-use purposes. The alleged Chinese government espionage activities against the African Union (AU) offer an initial example. The Chinese government funded the development of the AU headquarters in Addis Ababa, Ethiopia, and used a Chinese state-owned company to build it. The headquarters also used Huawei equipment for its telecom infrastructure. The AU alleged that, from 2012 to 2017, data were transferred to servers in Shanghai, China, each night between 12 a.m. and 2 a.m.[12] Huawei equipment potentially offered an access point to reroute traffic from AU headquarters to Shanghai. While China denies these allegations, the AU incident suggests how the CCP can leverage Chinese commercial companies for security purposes. China continues to invest in DI throughout Africa, indicating a potential broader interest in establishing an overseas presence through DI.

Many also view China's physical infrastructure and DI investments across Africa as a means for broadscale espionage. In Africa, Chinese companies have built roughly 186 government buildings and 14 sensitive intra-

---

[10] Poling, 2020, p. 8; However, not all China scholars agree on China's interest in Gwadar as a PLAN naval installation. For example, Daniel Kostecka suggested China's interest in using foreign airfields and ports as *places* rather than *bases*, with *places* focusing on periodic access to locations and supplies rather than permanent installations. He further refuted the notion of China's interest in Gwadar as a PLAN logistics port by indicating that it has limited military utility because of local instability, poor infrastructure and equipment, and vulnerability to air threats. For more details, see Daniel J. Kostecka, "Places and Bases: The Chinese Navy's Emerging Support Network in the Indian Ocean," *Naval War College Review*, Vol. 64, No. 1, Winter 2011.

[11] Poling, 2020, p. 6.

[12] John Aglionby, Emily Feng, and Yuan Yang, "African Union Accuses China of Hacking Headquarters," *Financial Times*, January 29, 2018.

governmental telecommunications networks and donated computers to 35 African countries.[13] Furthermore, from 2003 to 2017, a series of African countries signed $3.56 billion in loans with China for "policing, law and order, and dual use (civilian and military)" purposes.[14] The loans include DI for national security telecommunications, closed circuit television systems, and AI.[15] Chinese companies, in particular Huawei, support the development and operation of various telecom equipment and DI included in these loans.

Legal and policy changes in China that alter the CCP–private-sector relationship further suggest China's ability to leverage private-sector activities (including DI) for dual-use purposes. The CCP passed the National Intelligence Law in 2017, which requires all Chinese companies to comply with requests for assistance from the Ministry of State Security without any option for an appeal.[16] Therefore, any Chinese company that builds, supplies, or operates infrastructure must cooperate with the CCP, if requested. In August 2021, the Chinese government passed a new data-security law that all internet companies must comply with to cut down on the growing black market for consumer data used for identity theft and other crimes in China. However, the law still permits the Chinese government access to the data.[17] Viewed in the context of China's many infrastructure and investment initiatives, these projects offer the Chinese government and the PLA opportunities to access digital and physical infrastructure for noncommercial purposes. By extension, these noncommercial purposes could include elements of military power projection. We next discuss some potential nontraditional power-projection endeavors enabled by DI, beginning with a discussion of the Digital Silk Road (DSR).

---

[13] Joshua Meservey, "Government Buildings in Africa Are a Likely Vector for Chinese Spying," backgrounder, Washington, D.C.: Heritage Foundation, May 20, 2020.

[14] Nantulya, 2020, p. 7.

[15] Nantulya, 2020, p. 7.

[16] Lewis, 2019.

[17] Eva Xiao, "China Passes One of the World's Strictest Data-Privacy Laws," *Wall Street Journal*, August 20, 2021.

In 2015, China announced the DSR, an initiative tied to the BRI but focused entirely on digital technology, infrastructure, and services. As with the BRI, the DSR uses memorandums of understanding (MOUs) with governments to establish infrastructure investments.[18] The DSR areas of focus include

- investment in national telecommunications networks (fiber-optic cables, 5G and next-generation wireless networks, satellite-tracking ground stations)
- deployment of Chinese technology services globally (smart cities, security information systems, data centers).

Both pertain to DI and offer various means of creating and leveraging digital presence in competition and conflict.[19]

Furthermore, the DSR's scope appears more expansive than the MOUs. Although 16 DSR MOUs exist across Africa, Asia, Central and Eastern Europe, Latin America, and the Middle East, some note that China has current and planned DSR projects in over 130 countries.[20] In contrast to the BRI, which heavily leverages state-owned enterprises, the DSR relies on private-sector companies for its various projects.[21] The Chinese companies likely to support DSR projects include Huawei and ZTE for telecommunications networks and equipment; state-backed telecom providers (China Mobile, China Telecom, China Unicom); and key application providers, such as Alibaba, Tencent, and Baidu, for technology services.[22] As the 2017 National Intelligence Law indicates, although Chinese private-sector companies have fewer official ties to the CCP than state-owned enterprises, other mecha-

---

[18] Meia Nouwens, *China's Digital Silk Road: Integration into National IT Infrastructure and Wider Implications for Western Defence Industries*, London: International Institute for Strategic Studies, February 2021, pp. 7–9.

[19] Nouwens, 2021, pp. 7–9.

[20] Paul Triolo, Kevin Allison, Clarise Brown, and Kelsey Broderick, "The Digital Silk Road: Expanding China's Digital Footprint," Washington, D.C.: Eurasia Group, April 8, 2020, p. 2.

[21] Nouwens, 2021, pp. 7–9.

[22] Triolo et al., 2020, p. 1.

nisms exist for the government to leverage these entities. These mechanisms will reportedly still be in place after the new Chinese data-privacy law went into effect on November 1, 2021.[23] China's pursuit of the DSR suggests additional avenues for Beijing to project power globally.

Taken together, these efforts demonstrate not only that China wants to create and leverage a global power-projection capability but also that it may be pursuing nontraditional means to do so. The nontraditional means appear tied to DI. The next section explores a related assumption, DoD's growing reliance on DI building blocks and foundation elements.

Growing DoD Reliance on Digital Infrastructure

While the U.S. military already relies on DI to support its networks and communications, we expect this reliance to increase in the future because of military and technology trends. As a result, the future of DI and those who have significant OAC of it will have implications for U.S. defense and national security. We next outline how the U.S. military relies on DI today and why this matters for U.S. defense and national security.

Relationship Between Telecommunications and Military Networks

The U.S. military relies on DI for (1) data transport and (2) state-of-the-art networks and communication technology. DI provides transport communication infrastructure (a commercial transport layer) for DoD networks. Typically, DoD leases bandwidth from commercial companies to create capacity for data transport across military networks.[24] This includes leasing bandwidth for fiber-optic communications (both on land and undersea) and satellite communications.[25] Although most military networks use some

---

[23] Xiao, 2021.

[24] Theresa Hitchens, "Air Force to Launch 4G LTE at 20 More Bases Next Month," Breaking Defense website, November 12, 2020.

[25] U.S. Government Accountability Office, *DOD Management Approach and Processes Not Well-Suited to Support Development of Global Information Grid*, Washington, D.C., GAO-06-211, January 2006, p. 33.

commercial components, DoD does have dedicated, independent networks for certain activities. For example, some military satellite networks use DoD COMSATs, such as Wideband Global Satellite Communication, Advanced Extremely High Frequency, and Mobile User Objective System.[26]

To support and maintain a high-capacity transport layer—comprising hardware and software—DoD uses commercial telecom equipment. The DoD has a series of protections in place for networks that incorporate various commercial components and infrastructure. Although protections exist, such as encryption, the scale and scope of vulnerabilities associated with these networks may increase as DoD increases its reliance on DI and if an adversary can exploit the DI transport layer to copy, divert, or delete encrypted DoD message traffic.

The software components of the DI are also entirely based on commercial technologies. In the 1970s, DoD developed a software language, Ada, and associated tools, such as compilers, to enhance the security of software. However, the commercial sector never adopted Ada because defense contractors had difficulty finding software developers familiar with the language. Eventually, it was recommended that DoD abandon its Ada software requirements.[27] Today, the DI and DoD networks are based entirely commercially or academically developed software architectures.

The DoD also relies on DI for advanced and state-of-the-art communications technology. Innovation and development of advanced communications technology will continue to occur in the commercial realm to support the latest DI technologies, such as 5G wireless networking. We expect this trend to become increasingly relevant for DoD as the military moves toward exploiting 5G, new fiber-optic technologies, new COMSATs, and other new advanced communications technology.[28] As DoD builds or leverages state-

---

[26] Mike Gruss, "Pentagon: Narrowband? Wideband? Just Call Them Communications Satellites," webpage, SpaceNews, March 8, 2016.

[27] National Research Council, *Ada and Beyond: Software Policies for the Department of Defense*, Washington, D.C.: National Academy Press, 1997.

[28] Drew FitzGerald, "White House to Retool Pentagon Airwaves for 5G Networks," *Wall Street Journal*, August 10, 2020; Sydney J. Freedberg, "5G Experiments in US Pave Way to Battlefields Abroad," Breaking Defense website, November 4, 2020b.

of-the-art networks, they may increasingly rely on DI commercial components and supply chains.[29]

Finally, DoD relies on critical infrastructure, electric power grids, railroads, and water systems to support its forces at bases in the United States and overseas, all of which has become increasingly connected to DI (overseas and in the United States). The U.S. military also depends on critical infrastructure—ports, bridges, rail lines, etc.—to support its logistics networks and support functions. An adversary obtaining and using OAC of DI supporting critical infrastructure where U.S. forces were deployed could degrade military logistics and support to the forces.

Implications for DoD

The location and pace of advanced communication technology development, reliance of military networks on the global telecommunications network, and ability of China to potentially control different aspects of the global telecommunications network all pose significant areas of concern for DoD. First, the rate of advance of commercial communication technologies (e.g., 5G) is faster than that for military communication technologies. Second, U.S. and, perhaps even Chinese, military networks will likely rely to an even greater extent on the global telecommunications network in the future. Third, China may be able to exert greater control over the design and operation of the global telecommunications network in the future (China has funded the development of DI national champions, such as China Telecom, Huawei, and ZTE, perhaps not only for economic benefit).

---

[29] Important elements of DI supply chains include the foundries that manufacture state-of-the-art microchips. If these foundries are not trustworthy, they could provide microchips to DoD programs that contain hidden hardware or firmware compromises or vulnerabilities. DI technologies are, in many ways, driving and financially underwriting the development of the most advanced foundries in the world. For example, DI supply chains for state-of-the-art DI microchips provide the central processing units and high-speed multiband modems that go into 5G smartphones.

Ownership, Access, and Control of Digital Infrastructure Yield Benefits or Risks in Military Competition and Conflict

A country's degree of OAC of DI building blocks and foundational elements can yield tremendous benefits or risks in military competition and conflict. We hypothesize (1) that DI and foundational elements will significantly affect future warfare and, thus, (2) that an actor's OAC of DI, or lack thereof, will create advantages or disadvantages militarily. We first unpack this assumption by identifying shifting and future characteristics of warfare before applying revolution in military affairs (RMA) and military innovation processes to DI to understand what these changes mean for major military powers.

Characteristics of Future Warfare

Some characteristics of future warfare are likely shifting because of how DI is evolving, which may disrupt the U.S. way of war and traditional approaches to power projection. These changes will also affect future platforms and capabilities.

First, a country's physical footprint may no longer be sufficient to offer key military advantages in some scenarios. As noted in Chapter Two, digital presence can act as force multiplier in the absence of, or in addition to, a physical presence.

Second, similar to the first of our three assumptions, there is more than one way to project power beyond a country's borders through nontraditional means, many of which leverage DI.

Third, there is a growing rise and importance of commercially developed dual-use technology and capabilities that can be used for military and intelligence purposes. As a result, this technology will become an increasingly important aspect of future warfare. Examples include advanced communications that rely on commercial infrastructure (e.g., 5G) and the democratization of signals intelligence and ubiquity of open-source intelligence.[30]

---

[30] Freedberg, 2020b; Cortney Weinbaum, "The Intelligence Community's Deadly Bias Toward Classified Sources," Defense One website, April 9, 2021; Cortney Weinbaum,

Fourth, the proliferation of precision strike weapons will increase, and their size will continue to decrease because of the availability of high-performance chips.[31] China's focus on regional anti-access and area denial capabilities requires such weapons, e.g., the DF-21D antiship ballistic missile.[32] More specifically, the DF-17 relies on chips made by Taiwan Semiconductor Manufacturing Company (TSMC).[33] Finally, new electronic warfare (EW) and cyber capabilities will render command, control, and communication (C3) networks more vulnerable.[34] In 2018, it was reported that China had installed EW assets in the South China Sea to disrupt U.S. communications networks.[35]

Given these characteristics of future warfare and how they relate to DI, we next look more specifically at how DI may affect future warfare at the military system, operational concept, and organizational level.

Revolutions in Military Affairs and Military Innovation

Consistent with how military innovations and RMAs emerge, DI could potentially yield significant military opportunities for major powers. DI and foundational elements (microchips and technical standards) have a

Steven Berner, and Bruce McClintock, *SIGINT for Anyone: The Growing Availability of Signals Intelligence in the Public Domain*, Santa Monica, Calif.: RAND Corporation, PE-273-OSD, 2017.

[31] Ian Williams and Masao Dahlgren, "More Than Missiles: China Previews Its New Way of War," Washington, D.C.: Center for Strategic and International Studies, October 2019.

[32] Randy Huiss, *Proliferation of Precision Strike: Issues for Congress*, Washington, D.C.: Congressional Research Service, R42539, May 14, 2012, pp. 15–18.

[33] Ellen Nakashima and Gerry Shih, "China Builds Advanced Weapons Systems Using American Chip Technology," *Washington Post*, April 7, 2021.

[34] Ely Ratner, Daniel Kliman, Susanna V. Blume, Rush Doshi, Chris Dougherty, Richard Fontaine, Peter Harrell, Martijn Rasser, Elizabeth Rosenberg, Eric Sayers, Daleep Singh, Paul Scharre, Loren DeJonge Schulman, Neil Bhatiya, Ashley Feng, Joshua Fitt, Megan Lamberth, Kristine Lee, and Ainikki Riikonen, *Rising to the China Challenge: Renewing American Competitiveness in the Indo-Pacific*, Washington, D.C.: Center for a New American Security, December 2019, p. 14.

[35] Amanda Macias, "China Is Quietly Conducting Electronic Warfare Tests in the South China Sea," CNBC, July 5, 2018.

direct relationship to advanced military capabilities and have the potential to affect warfare in substantial ways. Moreover, both the United States and China have efforts underway to leverage these for military applications. Figure 3.1 shows how DI aligns with an approach for translating promising technologies into military systems, operational concepts, and organizational structures to achieve operational advantages.[36] We also rely on RMA and military innovation literature as they both outline the relationship between new technology and increased military effectiveness.[37]

Importantly, we do not claim that DI or semiconductors and technical standards will revolutionize warfare. Instead, we argue that DI and foundational elements have become more central to military systems and operational concepts and, therefore, will significantly affect warfighting in the future. Figure 3.1 illustrates how DI will likely continue to affect future warfare and provides evidence to support our contention where this is already occurring.

As shown in the figure, changes in the character and conduct of warfare often begin with technological change.[38] We frame each box as a question to understand whether and to what extent we observe DI-related changes occurring. We view a major technological change as driving the development of new military systems, operational concepts, and organizational adaptations.

Military Systems

DI building blocks and foundational elements underpin the development of many current and planned military technologies. In military innovation and RMA literature, technological change first translates into military systems. We see evidence of this today, particularly with microchips and advanced weapon systems. The development and operation of many

---

[36] Andrew F. Krepinevich, *The Military-Technical Revolution: A Preliminary Assessment*, Washington, D.C.: Office of Net Assessment, July 1992; Williamson R. Murray and Allan R. Millett, eds., Military Innovation in the Interwar Period, Cambridge, UK: Cambridge University Press, 1996.

[37] Krepinevich, 1992, p. 3; Adam Grissom, "The Future of Military Innovation Studies," *Journal of Strategic Studies*, Vol. 29, No. 5, October 2006.

[38] Krepinevich, 1992, p. 3.

FIGURE 3.1

Leveraging Technology to Achieve Significant Operational Advantages

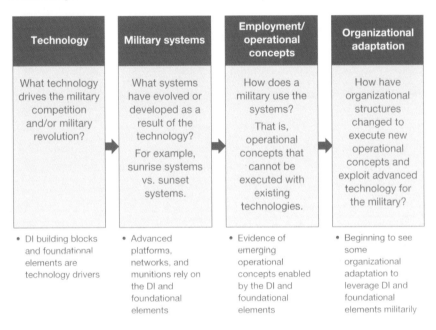

Technology	Military systems	Employment/operational concepts	Organizational adaptation
What technology drives the military competition and/or military revolution?	What systems have evolved or developed as a result of the technology? For example, sunrise systems vs. sunset systems.	How does a military use the systems? That is, operational concepts that cannot be executed with existing technologies.	How have organizational structures changed to execute new operational concepts and exploit advanced technology for the military?
• DI building blocks and foundational elements are technology drivers	• Advanced platforms, networks, and munitions rely on the DI and foundational elements	• Evidence of emerging operational concepts enabled by the DI and foundational elements	• Beginning to see some organizational adaptation to leverage DI and foundational elements militarily

SOURCE: RAND analysis informed by Krepinevich, 1992, and Murray and Millet, 1996.

advanced weapon systems, such as precision-guided munitions, rely on advanced microchips.[39] The U.S. DoD Research and Engineering office emphasized microelectronics as its top strategic priority in 2020, reflecting the importance of DI foundational elements for DoD military systems.[40]

---

[39] Nakashima and Shih, 2021; Andrew Eversden, "Pentagon, Intel Partner to Make More US Microchips for Military," webpage, C4ISRnet, March 19, 2021; Peter W. Singer, "The Lessons of World War 3," statement prepared for the U.S. Senate Committee on Armed Services, hearing on "Future of Warfare," 114th Cong., 1st Sess., Senate Hearing 114-211, Washington, D.C., November 3, 2015.

[40] Jason Sherman, "Microelectronics Is DOD's New No. 1 Technology Priority, Bumping Hypersonics to No. 3," *Inside Defense*, June 29, 2020.

Operational Concepts

Figure 3.1 illustrates how technological change becomes incorporated into operational concepts. Militaries develop new operational concepts—including new tactics and doctrine—that leverage the new or upgraded military systems that incorporate new, advanced technologies in place of existing technologies. In short, "operational concepts emerge that cannot be implemented" without aspects of DI building blocks and foundational elements to achieve greater operational effectiveness and advantages than what was possible before.[41] We see this today with current and emerging DoD and PLA operational concepts that leverage DI. For the PLA, DI building blocks and foundational elements underpin emerging and current operational concepts. Such Chinese concepts as *informatization* (ongoing), *intelligentization* (emerging), and *target-centric warfare* seek to leverage IT systems, information, and networks to create operational advantage.[42] On the other hand, DI building blocks serve as enablers of current and emerging DoD operational concepts, specifically Joint All-Domain Command and Control (JADC2),[43] which has been described as "a meta-network that can seamlessly share intelligence, orders, and other data across forces operating in all five military 'domains'—land, sea, air, space, and cyberspace."[44] Fur-

---

[41] Krepinevich, 1992, p. 19.

[42] Edmund J. Burke, Kristen Gunness, Cortez A. Cooper III, and Mark Cozad, *People's Liberation Army Operational Concepts*, Santa Monica, Calif.: RAND Corporation, RR-A394-1, 2020.

[43] The DoD has focused on using DI capabilities as part of an operational concept before, first with network-centric warfare in the late 1990s and early 2000s. We do not discuss network-centric warfare in depth here because current operational concepts, such as JADC2, have superseded it and because a broad historical analysis is beyond the scope of this research. See David S. Alberts, John J. Garstka, and Frederick P. Stein, *Network Centric Warfare: Developing and Leveraging Information Superiority*, 2nd ed., Washington, D.C.: U.S. Department of Defense, 1999.

[44] Sydney J. Freedberg, Jr., "Making War a Software Problem: JAIC Director on JADC2," Breaking Defense website, September 8, 2020a. See also, Jim Garamone, "Joint All-Domain Command, Control Framework Belongs to Warfighters," press release, November 30, 2020; David Vergun, "DoD Looking for Advanced Command, Control Solution," press release, Washington, D.C.: U.S. Department of Defense, June 4, 2021.

thermore, DI enables DoD information operations and information warfare focused on collecting and disseminating information about an adversary.

Organizational Adaptation

Finally, militaries adapt organizationally to execute operational concepts reliant on a given technological innovation. While more nascent than changes occurring in military systems and operational concepts, organizational restructuring to exploit new military systems and operational concepts tied to DI has started to occur. In 2015, the PLA restructured and created the Strategic Support Force, responsible for domains and systems that rely on DI building blocks and foundational blocks (information warfare, cyber, EW, space).[45] For the United States, the last major reorganization of DoD arguably occurred in 1986, with the Goldwater-Nichols Act. However, the military services and combatant commands have made some changes since then, such as elevating U.S. Cyber Command as a unified combatant command and creating the Space Force in 2019. These changes indicate DoD's recognition of space and cyber as strategically important domains.[46]

Past work on RMAs and military innovation indicate that such changes underway in military systems, operational concepts, and some organizational restructuring tied to DI may affect the character and conduct of warfare.[47] As a result, a country's control of and access to DI and access to DI foundational elements, especially trustworthy microchips, will become increasingly important for achieving the desired level of military effectiveness resulting from those changes.

---

[45] John Costello and Joe McReynolds, *China's Strategic Support Force: A Force for a New Era*, Washington, D.C.: National Defense University Press, 2018, p. 1.

[46] Public Law 116-92, National Defense Authorization Act for Fiscal Year 2020, Subtitle D, United States Space Force Act, December 20, 2019; Jim Garamone and Lisa Ferdinando, "DoD Initiates Process to Elevate U.S. Cyber Command to Unified Combatant Command," press release, Washington, D.C.: U.S. Department of Defense, August 18, 2017.

[47] We do not define the scale and scope of this change beyond identifying the shift taking place.

Summary

These assumptions remain foundational to our research in two ways. First, we have hypothesized that China's use of DI goes beyond economic and domestic purposes to military and security application on a global scale. Second, given that DoD will become more reliant on DI and that DI will likely affect future warfare significantly, the United States should seek to shape and control it rather than be beholden to an adversary nation that does. In particular, the United States should better understand the emerging DI competition because of its implications for national security and defense. The next section builds on these assumptions to present a characterization of DI competition between the United States and China.

How to Think About the Competition for Digital Infrastructure

The DI competition underway today between the United States and China emerged in the 2000s and sits within the broader U.S.-China competition for status, influence, and power (military, economic, etc.).[48] While traditional elements of power still shape competition and create competitive advantages, we see evidence of DI playing a growing role in creating advantage in the economic, military, information, and diplomatic dimensions of power. To understand how these dimensions affect the broader competition, we next provide a brief methodological overview and discussion of power.

Dimensions of Power

Our analysis characterizes national power using four dimensions widely cited as contributors to competition outcomes: military, economic, informa-

---

[48] For more information on the role of influence in the strategic competition, see Michael J. Mazarr, Bryan Frederick, John J. Drennan, Emily Ellinger, Kelly Eusebi, Bryan Rooney, Andrew Stravers, and Emily Yoder, *Understanding Influence in the Strategic Competition with China*, Santa Monica, Calif.: RAND Corporation, RR-A290-1, 2021.

tion, and diplomatic.[49] States often use a combination of these four means to compete and create relative advantage vis-à-vis other actors.[50] The 2018 National Defense Strategy refers to national power as a subset of various factors, including the four dimensions noted earlier.[51] The 2017 National Security Strategy also discusses national power as an integration of political, economic, and military elements.[52] Our four dimensions of power remain consistent with these strategic guidance documents. Furthermore, we define competition broadly as "the attempt to gain advantage, often relative to others believed to pose a challenge or threat, through the self-interested pursuit of contested goods such as power, security, wealth, influence, and status."[53]

For each dimension of power, a set of factors (power centers) offers opportunities to create competitive advantage. For example, in the 1980s, oil was arguably the power center for economic power; the more oil a country had meant the greater its economic advantage. Power centers shift over time with changes in geopolitics, economic trends, technology, etc. Today, for each of the four dimensions, the power centers appear to be moving toward DI. We hypothesize that these trends will become even more pronounced in the future.

---

[49] For a more in-depth discussion of competition, see Michael J. Mazarr, Jonathan Blake, Abigail Casey, Tim McDonald, Stephanie Pezard, and Michael Spirtas, *Understanding the Emerging Era of International Competition: Theoretical and Historical Perspectives*, Santa Monica, Calif.: RAND Corporation, RR-2726-AF, 2018; for a more in-depth discussion of power in international relations, see David A. Baldwin, "Power and International Relations," in Walter Carlsnaes, Thomas Risse, and Beth A. Simmons, eds., *Handbook of International Relations*, Newbury Park, Calif.: SAGE Publications, 2013.

[50] Mazarr, Blake, et al., 2018, p. 19; Kenneth N. Waltz, *Theory of International Politics*, reprint, Long Grove, Ill.: Waveland Press, Inc., 2010, p. 98.

[51] DoD, *Summary of the 2018 National Defense Strategy of the United States of America: Sharpening the American Military's Competitive Edge*, Washington, D.C., 2018, p. 4.

[52] White House, *National Security Strategy of the United States of America*, Washington, D.C., December 2017, p. 26.

[53] Mazarr, Blake, et al., 2018, p. 5.

The Digital Infrastructure's Effect on Dimensions of Power

The growing influence of DI on the military, economic, information, and diplomatic dimensions of power suggests that DI may play a greater role in shaping the U.S.-China competition.

First, trends in military technology, organization, and strategy rely on and leverage DI. Advanced weapon systems require microchips; operational concepts center on networks enabled by DI; and overseas posture relies on communications enabled by DI building blocks. PLA operational and doctrinal concepts, such as target-centric warfare and informatization, stress the importance of networks, information, and information systems to warfare.[54] Additionally, the development of advanced PLA platforms, such as hypersonic missiles, relies on a DI foundational element (microchips).[55]

Similarly, advanced U.S. weapon platforms also rely on microchips and, more broadly, on the DI for development and operation.[56] Furthermore, an emerging U.S. operational concept, JADC2, leverages integrated networks to create military advantage and create multiple targeting dilemmas for adversaries by linking sensors and shooters more seamlessly across the joint force.[57] U.S. Deputy Secretary of Defense Kathleen Hicks recently issued a memo describing data as a strategic asset, "critical to improving performance and creating decision advantage at all echelons from the battlespace to the board room, ensuring U.S. competitive advantage."[58] DI consists of networks that transmit, generate, and store data that the deputy secretary refers to as strategically important.

Second, DI companies and technologies increasingly represent a greater share of economic activity, not just in the United States but globally. The importance of DI to economic growth was recognized in the United States

---

[54] Burke et al., 2020.

[55] Nakashima and Shih, 2021.

[56] Singer, 2015.

[57] Garamone, 2020; Vergun, 2021.

[58] Kathleen Hicks, "Creating Data Advantage," memorandum for Senior Pentagon Leadership, Commanders of the Combatant Commands, Defense Agency and DoD Field Activity Directors, Washington, D.C.: U.S. Department of Defense, May 5, 2021.

early on. In 1994, Vice President Al Gore predicted that DI would have major economic benefits for the United States and other countries. At the time, the Clinton administration supported the development of the internet and advanced telecommunications networks because of their perceived economic benefits and promoted the adoption of these networks in the international community.[59] Most recently, in the past five years, Big Tech companies—Facebook, Alphabet,[60] Amazon, Microsoft, Apple—have doubled their share of the overall capitalization of the U.S. stock market. As of 2021, these companies account for 25 percent of the value of all companies in the Standard and Poor's 500, with a combined value of $8 trillion.[61] These companies contribute to DI building blocks and rely on DI and foundational elements. Facebook and Google already invest in and own submarine cables.[62] Apple now designs its own microchips, although they are manufactured in Taiwan.[63] These trends suggest that DI companies—such as Facebook, Apple, Amazon, Netflix, and Google (the so-called FAANG stocks)—will contribute significantly to economic growth in the future. The chief executive officer (CEO) of Microsoft recently stated that "[o]ver a year into the pandemic, digital adoption curves aren't slowing down. They're accelerating,"[64] These trends also indicate that DI companies offer levers to create economic growth and advantage.

Third, DI underrides the transmission, exploitation, and use of information for intelligence operations and broader information operations.[65] Societies across the globe could not communicate without fiber-optic cables (on

---

[59] Al Gore, remarks, Inauguration of the First World Telecommunication Development Conference (WTDC-94), International Telecommunication Union, Buenos Aires, March 21, 1994..

[60] The parent company of Google.

[61] "Five Tech Giants Just Keep Growing," *Wall Street Journal*, May 1, 2021.

[62] Alex Webb, "A Few Glass Strands Will Protect U.S. Tech From China," Bloomberg, April 5, 2021.

[63] Mark Gurman, Debby Wu, and Ian King, "Apple Aims to Sell Macs with Its Own Chips Starting in 2021," Bloomberg, April 23, 2020.

[64] "Five Tech Giants . . . ," 2021.

[65] We discuss the information dimension in the context of intelligence operations and information operations to reduce overlap with the military dimension. However, we

land and undersea), cellular networks and infrastructure, the internet, satellites, and foundational elements essential to all DI building blocks (microchips and standards). As bandwidth needs increase—people and systems require more bandwidth to process, store, and transmit data—DI will likely become even more important to the information dimension. Submarine cable demand, for example, has surged in recent years because of the data requirements of streaming services, such as Netflix.[66] Control of and access to DI will offer more opportunities to create advantage in the information dimension of power.

Fourth, DI underpins many diplomatic efforts to build multilateral and bilateral relationships and plays a greater role in international institutions. DI, given its widespread use and ubiquity for many aspects of society and a country's elements of national power (military, economic, information), offers a growing form of influence globally. From China's BRI and DSR to the U.S. infrastructure plan that includes efforts to secure U.S. supply chains, countries are pursuing national strategic efforts to ensure and grow advantages in DI, given its importance for national power and national security.

In the next section, we build on these trends to offer an initial description of the U.S.-China DI competition, how to think about it, and how each country appears to be competing.

The U.S.-China Competition for Digital Infrastructure

A subset of the broader U.S.-China competition relates to DI and traces back to the 1990s and 2000s, when China started producing DI companies and technologies. By the 2000s, China also introduced high-level strategic initiatives leveraging DI. China's interest in DI, seen through state-led initiatives, PLA concepts, and the use of private-sector DI companies, resulted in an increased PRC presence within DI that began to challenge the market presence of the United States and its allies and partners. For many years,

---

recognize that information remains an essential facet of all military operations; thus, this trend could apply to the military dimension as well.

[66] Jeff Hecht, "Submarine Cable Goes for Record: 144,000 Gigabits from Hong Kong to L.A. in 1 Second," *IEEE Spectrum*, January 3, 2018.

the United States and its allies maintained DI leadership positions. China's emergence and strong interest in DI led to what we term the *U.S.-China DI competition.*

We hypothesize that, similar to how the United States views overseas military presence as a strategic means to compete and deter, the PRC views DI as a strategic means for competition.[67] Importantly, neither overseas military presence nor DI encapsulates the full spectrum of capabilities that the United States and China use to compete. However, we use the comparison to emphasize why DI competition rests within the long-term U.S-China strategic competition.

Despite DI's effect on various dimensions of power, the United States and China approach it differently. China's strategy for achieving long-term strategic goals appears focused heavily on using and exploiting DI, while the U.S. power stems from a post–World War II (WWII) model founded on institutions, power projection, and a rules-based liberal order.[68] This divergence in approach creates asymmetries in competitive strategies, means, and mindsets.

Grand Strategies

U.S. and PRC grand strategies differ in terms of approach, means, motivation, and timeline but remain similar in their shared goal of global status and influence.[69] U.S. grand strategy largely centers on the post-WWII liberal order, alliance structures, and military power and presence. The U.S. commitment to uphold, enforce, and support the international order remains a

---

[67] We have not verified that Chinese writings support this hypothesis and recognize that this hypothesis is based on the research of U.S. China scholars (e.g., Burke et al., 2020) and on our analysis of open-source research on Chinese initiatives (e.g., DSR, BRI, Made in China 2025). Additionally, we based this assumption on several of China's actions in recent years, using DI for espionage purposes and investment in DI in foreign countries.

[68] For more details on the U.S.-led liberal order, see Michael J. Mazarr, *Summary of the Building a Sustainable International Order Project*, Santa Monica, Calif.: RAND Corporation, RR-2397-OSD, 2018.

[69] Grand strategies shape a country's approach and role as a global power; concepts, capabilities, and organizational structures implement a grand strategy.

bedrock of U.S. grand strategy since 1945.[70] In contrast, China's grand strategy has changed more since WWII. Recent RAND research identified China's four grand strategies since 1949. What China scholars define as China's first two grand strategies focused inward on recovering and reforming the country domestically.[71] By the 1990s, however, China's grand strategic focus shifted outward toward building national power. The shift reflected China's perception of the United States as an existential threat following the end of the Cold War. The most current grand strategy, termed *national rejuvenation*, focuses on President Xi Jinping's 2050 goals of becoming a global power and achieving the "China Dream."[72]

Similar to grand strategies, U.S. and PRC strategies and motivations for competition differ. The United States seeks to maintain the international system and its role in the world through military strength, international organizations, partners and allies, and economic strength (all increasingly reliant on DI). On the other hand, the PRC likely desires both (1) a stronger position within and (2) a revision of the international system.[73] Beijing seeks status, and advantage, through economic, informational, military, and geopolitical means (many centered on DI).

---

[70] Hal Brands, *American Grand Strategy and the Liberal Order: Continuity, Change, and Options for the Future*, Santa Monica, Calif.: RAND Corporation, PE-209-OSD, 2016, pp. 1–2; We define *international order* as "the body of rules, norms, and institutions that govern relations among the key players in the international environment" (Mazarr, 2018, p. 3). See also Michael J. Green, *By More Than Providence: Grand Strategy and American Power in the Asia Pacific Since 1783*, New York: Columbia University Press, 2017.

[71] Andrew Scobell, Edmund J. Burke, Cortez A. Cooper III, Sale Lilly, Chad J. R. Ohlandt, Eric Warner, and J. D. Williams, *China's Grand Strategy: Trends, Trajectories, and Long-Term Competition*, Santa Monica, Calif.: RAND Corporation, RR-2798-A, 2020, pp. 5–21.

[72] Scobell et al., 2020, pp. 5–21.

[73] Office of the Secretary of Defense, 2020, p. v; For more discussion of China's goals with respect to the international system, see the discussion of China's 19th Party Congress in Michael J. Mazarr, Timothy R. Heath, and Astrid Stuth Cevallos, *China and the International Order*, Santa Monica, Calif.: RAND Corporation, RR-2423-OSD, 2018, pp. 18–20.

Strategic Efforts

We surveyed and identified several U.S. and PRC key strategic efforts to understand the foundations of each country's competitive approach and how it relates to DI. We leveraged previous RAND analyses on these efforts to create Tables 3.2 and 3.3 and do not present any original research therein. We instead chose efforts foundational to U.S. and PRC competitive strategies, means, and mindsets and, when possible, identified those most pertinent to DI. The subsequent discussion does not intend to be comprehensive but rather to be illustrative of some broader trends in U.S. and PRC approaches to DI.

The U.S. approach to creating, sustaining, and exercising power began in the post-WWII era and uses long-standing institutions (see Table 3.2).[74] These institutions use alliances, financial loans, and security and trade agreements (in addition to others) as the foundation for bilateral and multilateral relationships globally. Given that many of these efforts, organizations, and international institutions date back to the 1940s, they do not incorporate DI in any official or systematic way.

However, the U.S. government has signaled strategic interest in DI intermittently over time and much more frequently in the past five years. From a national security lens, we see this in DoD's 5G initiative (initiative focused on wireless infrastructure and technology), the Joint Artificial Intelligence Center (uses data from DI building blocks and networks within DI to process data), the establishment of the Space Force (organizational change focused on space), and the elevation of U.S. Cyber Command to a unified combatant command. Most recently, the Infrastructure Investment and Jobs Act (i.e., Biden infrastructure plan) further indicates that the U.S. government views the DI and its foundational elements as strategically important for the country.[75] Part of the new law focuses on reshoring state-of-the-art microchip

---

[74] The table does not include the United Nations, although we acknowledge its importance to the U.S. approach to global power and influence.

[75] White House, "President Biden's Bipartisan Infrastructure Law," press release, November 2021.

TABLE 3.2
Key U.S.-Led Strategic Efforts

Type	Effort	Origin	Description	DI link
National strategic objective (initiative)	Biden infrastructure plan	2021	Legislation for U.S. infrastructure plan	Direct
International organization (alliance structure)	World Bank	1944	International financial institution	Indirect
	International Monetary Fund	1945	International financial institution	Indirect
	North Atlantic Treaty Organization	1949	Intergovernmental military alliance with multiple country members	Indirect
Defense strategy (concept)	Military Power Projection[a]	Post-WWII	Military strategy for projecting military power beyond a country's borders	Indirect
	2018 National Defense Strategy[b]	2018	Strategic military guidance for DoD	Indirect
	JADC2[c]	2018	Operational concept that integrates networks across domains	Direct
Defense organization	Unified Combatant Commands[d]	1986	Geographic and functional military commands	Indirect
	U.S. Cyber Command	2018	U.S. Cyber Command elevated to a unified combatant command	Direct
	U.S. Space Force	2019	Department of the Air Force service established for space operations	Direct

Table 3.2—Continued

Type	Effort	Origin	Description	DI link
	DoD Joint Artificial Intelligence Center	2018	Center for leveraging AI expertise for DoD	Direct
Defense initiative	U.S. Defense Industrial Base[e]	Post-WWII	Companies that provide support to DoD	Direct
	DoD 5G Initiative[f]	2018	Initiative to develop 5G and next-generation capabilities for warfighters	Direct

[a] David Ochmanek, *Restoring U.S. Power Projection Capabilities: Responding to the 2018 National Defense Strategy*, Santa Monica, Calif.: RAND Corporation, PE-260-AF, 2018.

[b] DoD, 2018.

[c] John R. Hoehn, "Joint All-Domain Command and Control (JADC2)," Washington, D.C.: Congressional Research Service, IF11493, June 4, 2021.

[d] Pub. L. 99-433, Goldwater-Nichols Department of Defense Reorganization Act of 1986, October 1, 1986.

[e] Heidi M. Peters, "Defense Primer: U.S. Defense Industrial Base," Washington, D.C.: Congressional Research Service, IF10548, January 22, 2021.

[f] DoD, *Department of Defense 5G Strategy Implementation Plan: Advancing 5G Technology & Applications Securing 5G Capabilities*, Washington, D.C., December 17, 2020, p. 3.

TABLE 3.3
Key Chinese Strategic Efforts

Type	Effort	Origin	Description	DI link
National strategic objective (initiative)	Military-Civil Fusion[a]	1990s	Strategy to achieve military and economic modernization in support of 2049 National Rejuvenation goals	Direct
	China Dream[b]	2012	PRC strategic end state: "intention to realize the country's revitalization as a great power by midcentury"	Indirect
	BRI	2013	Chinese global infrastructure plan	Direct
	Made in China 2025[c]	2015	Industrial plan to reduce PRC reliance on foreign firms for emerging and advanced technology	Direct
	DSR[d]	2015	Tied to the BRI; focused on digital technology, infrastructure, and services	Direct
Economic organization	China Development Bank	1994	Government development bank	Indirect
Defense strategy (concept)	Anti-access and area denial strategy[e]	1990s	Strategy to prevent U.S. forces from operating in PRC combat theater	Indirect
	Informatization[f]	2004	Emphasizes information as an instrument and condition for warfighting	Direct
	Target-Centric Warfare[g]	2015	"[C]oncept of attacking critical points in the enemy's operational system to achieve decisive effects with minimal collateral damage"	Direct
	Intelligentization[h]	~2016	Emphasis on big data and AI to create military advantage	Direct
Defense organization	Strategic Support Force[i]	2015	PLA organization for information warfare (integration of cyber, EW, space)	Direct

48

Table 3.3—Continued

Type	Effort	Origin	Description	DI link
Defense initiative	Defense Industrial Base[j]	1980s	PRC companies that provide support to PLA	Indirect

[a] Young Professionals in Foreign Policy, *US-China Futures: Briefing Book*, New York: Schmidt Futures, March 2021, p. 42; Office of the Secretary of Defense, 2020, pp. v–vi.

[b] Timothy R. Heath, Derek Grossman, and Asha Clark, *China's Quest for Global Primacy: An Analysis of Chinese International and Defense Strategies to Outcompete the United States*, Santa Monica, Calif.: RAND Corporation, RR-A447-1, 2021, p. 17.

[c] Karen M. Sutter, "Made in China 2025' Industrial Policies: Issues for Congress," Washington, D.C.: Congressional Research Service, IF10964, August 11, 2020.

[d] Nouwens, 2021.

[e] Roger Cliff, Mark Burles, Michael S. Chase, Derek Eaton, and Kevin L. Pollpeter, *Entering the Dragon's Lair: Chinese Antiaccess Strategies and Their Implications for the United States*, Santa Monica, Calif.: RAND Corporation, MG-524-AF, 2007.

[f] Burke et al., 2020, pp. 6–7.

[g] Burke et al., 2020, p. 15.

[h] Burke et al., 2020, pp. 21–22.

[i] Kevin L. Pollpeter, Michael S. Chase, and Eric Heginbotham, *The Creation of the PLA Strategic Support Force and Its Implications for Chinese Military Space Operations*, Santa Monica, Calif.: RAND Corporation, RR-2058-AF, 2017, pp. 13–14.

[j] DoD, 2020, pp. 143–148.

foundries to the United States and expanding broadband communications to rural and disadvantaged U.S. communities.[76]

Independent of these efforts, the United States has also developed alternatives to China's infrastructure investment initiatives that focus almost entirely on aspects of DI. We characterize these U.S. initiatives more as tit for tat than as foundational to the U.S. approach to competition. For example, the United States founded the International Development Finance Corporation (DFC) in 2019 as "an investment tool we [United States] have to compete."[77] The DFC attempted to counter Chinese investment efforts in Greece (purchasing a shipyard) and Ethiopia (offering assistance for 5G) directly by offering an alternative investment portfolio. Additionally, in 2019, the DFC introduced a lending initiative to a U.S. company to build a submarine fiber-optic cable connecting the United States, Singapore, Indonesia, and Palau as an alternative to "Huawei-built undersea networks."[78] Furthermore, the United States has also been a part of multilateral initiatives to offer alternatives. The Group of Seven—which includes the United States—recently introduced Build Back Better World (B3W), an initiative designed to "unleash hundreds of billions of dollars for projects in needier countries . . . as an explicit alternative to Chinese infrastructure offerings."[79] Although the United States does consider DI on a strategic level, it does not appear to incorporate DI as an instrument for competition.

In contrast, Table 3.3 shows China's systematic use of DI across various dimensions of power. In many cases, the PRC attempts to integrate activities typically siloed to a particular sector for DI initiatives, as seen with the Military-Civil Fusion strategy, Made in China 2025, and DSR. We use *systematic* to refer to China's use of and interest in DI across multiple dimensions of power. Many of China's cross-cutting strategic efforts use DI as the basis, or currency, for creating bilateral and multilateral relationships (e.g., DSR).

---

[76] Luis Melgar and Ana Rivas, "Biden's Infrastructure Plan Visualized: How the $2.3 Trillion Would Be Allocated," *Wall Street Journal*, April 1, 2021.

[77] Quotation from Rep. Michael McFaul, quoted in Stu Woo and Daniel Michaels, "China Buys Friends with Ports and Roads. Now the U.S. Is Trying to Compete," *Wall Street Journal*, July 15, 2021.

[78] Woo and Michaels, 2021.

[79] Woo and Michaels, 2021.

China's involvement in DI began decades after U.S. efforts did; however, since the 1980s and 1990s, China has increasingly resourced and organized to use DI for influence and advantage. For example, the PLA's introduction of informatized wars in 2004 and other military concepts stressed the critical role networks and information systems play in creating military advantage.[80]

Summary

We identified two strategic asymmetries from our analysis: (1) The United States seeks to maintain power whereas China aims to grow its power, and (2) China and the United States both rely on and leverage DI, but their approaches for leveraging DI and the degree to which they use it differ. Beijing appears to view DI as central to its competitive strategies and means; the United States perceives it as an enabler, recognizing distinct parts of DI as important for achieving strategic goals.

Additionally, the United States and China have different timelines for their respective achievement and maintenance of global power status. U.S. scholars on China's grand strategy have noted that, after the Cold War, Beijing perceived the United States as a growing threat, presenting soft- and hard-power challenges to China.[81] These scholars have also noted that this perception grew over time, particularly in the 2000s. This perception of the United States as the sole superpower informed the development of China's third grand strategy to create national power and the most recent strategy, focused on national rejuvenation and achieving the China Dream.[82] Perhaps most important, we hypothesize that this asymmetry showcases a U.S. competitive advantage: the United States built the international order that China likely seeks to lead.[83] While research on China's approach to and goals for

---

[80] Burke et al., 2020.

[81] Scobell et al., 2020, p. 22.

[82] Scobell et al., 2020, pp. 16–19.

[83] Heath, Grossman, and Clark, 2021, pp. xvi–xvii, 27, 42–53, 207; Michael Mazarr first identified this competitive advantage in Mazarr, 2018, p. 2.

competition remain unclear, this hypothesis stems from analysis of China scholars' deduction of Chinese writings and assumption of Chinese views.

Trends seem to indicate that DI will tie to a country's national power in the future and, by extension, to the resources available to compete. This, in turn, will shape the U.S-China strategic competition. While China has a series of ongoing strategic initiatives to leverage DI to achieve for the country's own benefit, this does not necessarily translate into a competitive advantage for China, given the structural factors that exist in the international system that favor the United States (as the leader of the liberal order, leading military, largest economy, etc.).

CHAPTER FOUR

Digital Infrastructure Evolution, Trends, and Asymmetries

In this chapter, we describe the evolution of DI from its earliest beginnings to the present (2021), including the activities of leading U.S., Chinese, and other foreign companies that have competed for market share and control over DI networks and their supply chains. We examine the history of the DI and the role of the United States and China in its development. We also examine the U.S. and PRC roles in DI building-block trends. Additionally, we look at two DI foundational elements—microchips and standards—as they affect the development and control of DI.

We first outline DI's evolution before identifying DI structural trends and asymmetries. We then discuss trends and asymmetries associated with DI foundational elements. We conclude with a historical and current technical assessment to better understand the U.S-China balance of power within DI and how this may have changed over time.

Digital Infrastructure Building Blocks

DI has evolved over time, as have the relative advantages of the United States and China in each area. In the 1970s, 1980s, and 1990s, U.S. companies, the U.S. National Science Foundation (NSF), and DoD led the development of all DI building blocks. Over time, technology leadership became more diffuse, and major parts of the U.S. industrial base for telecommunications and personal computing were sold off to foreign entities in a series of transactions. AT&T spun off its equipment and research businesses into a company later called Lucent. IBM sold its PC division to Lenovo in 2005.[1] In 2006 Lucent was sold to Alcatel.[2] Bell Labs was acquired by Alcatel-Lucent when the French company Alcatel acquired Lucent.[3] Motorola, at the time a major player in the cellular infrastructure market, sold its wireless networking equipment division to Nokia in 2010.[4] And, in 2010, Nokia acquired Alcatel-Lucent and, with it, Bell Labs, which was still located in New Jersey (it was operated as a separate subsidiary because of the sensitive communications technologies it worked on). By 2010, the significant surviving players in Asia (Samsung, Huawei, ZTE) and Europe (Ericsson, Nokia) emerged to dominate the wireless infrastructure market. Also, in the same period, AT&T and Lucent retreated from the submarine cable business, leaving that market to European and Asian companies. Only in the space and terrestrial DI network segments did U.S. companies remain significant market players. The U.S. companies Cisco Systems and Juniper Networks remain major players in the terrestrial switching and routing market, while Boeing, Space Systems Loral (now Maxar Technologies), and SpaceX are major players in the COMSAT market. Figure 4.1 depicts this evolution and highlights key events in the evolution of DI. The box colors indicate the role of the United States or China in each event: red for China and blue for the United States. If the United States was the lead actor, we code the text or box in blue and do the same, in red, when China leads.

The origins of DI predate events shown in Figure 4.1. Events that led to the invention of the internet, the first mobile phone, and the first intercontinental digital fiber-optic cable highlight important trends and elicit key inflection points. As a result, we begin our historical discussion with the origins of major DI milestones, such as the first communication networks. We walk through DI's evolution holistically rather than breaking out each building block individually. We do so because DI functions as an integrated

---

[1] Lenovo, "Lenovo Marks Decade of Success Since Acquisition of IBM's PC Business," press release, April 30, 2015.

[2] Keith Griffith, "Nokia to Acquire Alcatel-Lucent for $16.6 Billion," SDxCentral, April 15, 2015.

[3] Griffith, 2015.

[4] Jenna Wortham, "Nokia Siemens Agrees to Pay Cash for Division of Motorola," *New York Times*, July 19, 2010.

FIGURE 4.1

Key Events in the Evolution of Terrestrial, Wireless, and Submarine Networks

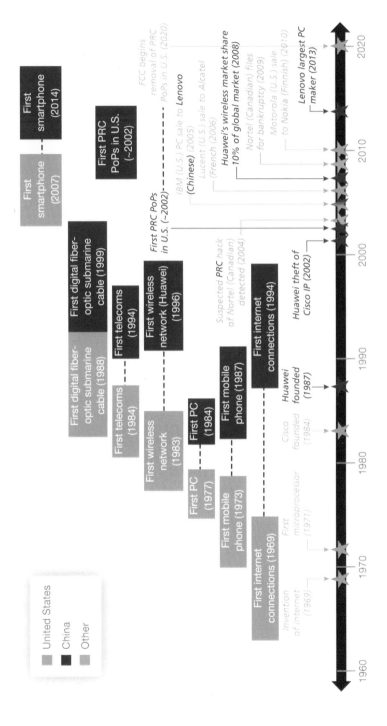

entity. DI comprises many networks, inventions, and companies and should be understood as the sum of its parts.

Origins

Telecommunications

The first communication networks were terrestrial networks that connected companies; cities; towns; and, eventually, the vast majority of homes using the public telephone switch network. The first terrestrial telecommunications networks for voice were deployed in the United States and Britain in the 1880s and relied on the invention of the telephone by Alexander Graham Bell in 1876 to transmit and receive voice calls. Eventually, one company developed a network capable of supporting reliable long-distance voice calls in the United States: American Telephone and Telegraph, whose corporate descendent is now known simply as AT&T.

AT&T became a private monopoly corporation and was officially sanctioned as a private monopoly by court order in 1913, after agreeing to allow local phone companies to connect to its network.[5] By 1915, the AT&T network had connected New York to San Francisco. Overseas networks evolved differently, as government-owned monopoly companies were tasked to build out the national networks in different countries.

In 1925, AT&T created its own research and development (R&D) laboratory, Bell Labs, which was responsible for many inventions and innovations important for the development of new computing and communication technologies. AT&T continued to dominate the U.S. telecom industry and, in 1984, was broken up in response to an antitrust lawsuit brought by the U.S. government. The core of AT&T remained in the parent company, the long-distance company, and its manufacturing arm, which made telephone and network equipment. However, the local regional networks, the so-called Baby Bells, were spawn spun off as separate corporate entities. In the breakup, AT&T retained Bell Labs. AT&T was no longer a monopoly in either the long-distance market, where Sprint and Microwave Communications Incorporated (MCI) emerged as competitors, or in the telephone equipment market, where a host of new companies emerged. We count this

---

[5] "AT&T's History of Invention and Breakups," *New York Times*, February 13, 2016.

key event, the breakup of AT&T in 1984, as the birth of the first modern competitive U.S. telecom carrier.[6]

Internet

The internet was created by a consortium of R&D organizations supported by funding from DoD. The first internet connections were established in 1969. From that moment, the internet grew, especially in the United States and Europe, first being used by academic institutions. In the United States, NSF, the Department of Energy, and the National Aeronautics and Space Administration (NASA) also played key roles in the 1980s and 1990s in the internet by maturing internet technologies and creating governance structures that ensured the interoperability and continued scalability of the internet. NSF nurtured the development of NSFnet, which went online in 1986 and was not decommissioned until 1998, when key elements of the internet were transferred to the commercial sector. It was during this period of NSF oversight that the internet experienced explosive growth and when physicists at CERN invented the World Wide Web protocol.[7]

Eventually, U.S. and European telecom companies were able to provide the transport infrastructure to support the internet using new equipment from such Silicon Valley companies as Cisco Systems. Initially, the internet was distinct and separate from the voice networks that telecom companies traditionally provided over closed networks. In contrast, the internet was based on open standards, and as internet technologies have continued to advance, new suppliers have emerged from Silicon Valley, such as Juniper Networks. These firms have provided the progressively more capable routing and switching equipment needed to support the growth in internet traffic. In the 1980s and 1990s, U.S. telecom carriers and their suppliers had a significant lead over the telecom sectors of other countries.

The internet was deployed first in the United States and, soon thereafter, in Europe. The U.S. government supported fundamental research into

---

[6] Another important aspect of telecommunications are the suppliers and supply chains that provide the components for these networks. Early in the development of the U.S. terrestrial network, AT&T made much of its own equipment and supplied itself with telecom switches and phones.

[7] NSF, "A Brief History of NSF and the Internet," fact sheet, August 13, 2003.

internet technologies but did not intervene in the new market for internet communications equipment. In the initial deployments of telephonic and internet terrestrial networks, U.S. carriers could choose to use equipment from anywhere in the world, although the most advanced equipment was available from U.S. manufacturers in many cases. This open supplier market approach in the United States continued to prevail for several decades and only recently has been limited, as we will discuss later.

Wireless

Wireless telecommunications networks developed even more rapidly than terrestrial networks. Cellular communications were developed after terrestrial wireline telecommunications had become well established. Cellular technologies were pioneered in the United States largely by private companies. Motorola developed and demonstrated the first mobile cellular phone and network in 1973,[8] more than two decades before the first Chinese mobile phone came to market. The first U.S. cellular network was established a decade after the first mobile phone, in 1983.[9] It took more than a decade for Huawei to deploy its first mobile phone network in China, in 1997.[10]

U.S. telecom carriers experienced rapid growth in network subscribers and revenues as the price of mobile phones fell, as service costs declined, and as network speeds and features increased. Traditional terrestrial network carriers realized that growth opportunities were much greater in wireless networks and started acquiring newly established wireless carriers. For example, AT&T acquired McCaw Cellular in 1994 and, over time, built out its own nationwide wireless network.

Submarine Cables

The first trans-Atlantic submarine cable was laid in 1858. It was an analog, low-data-rate system that permitted telegraph communications at 12 words per minute. It lasted only 26 days before it failed. By 1956, several technol-

---

[8] Motorola, *A Timeline Overview of Motorola History: 1928–2008*, 2008.

[9] Steel in the Air, Inc., "The Wireless Carrier Timeline & Industry Evolution: Pre-1983–Present," webpage, 2014.

[10] Yun Wen, *The Huawei Model: The Rise of China's Technology Giant*, Chicago: University of Illinois Press, 2020.

ogy advances were made to enable submarine cables to carry a small number of analog voice calls.[11] The digital technologies needed for high-capacity long-range fiber-optic submarine cables were invented and perfected in the United States at Bell Laboratories in New Jersey before they were combined to achieve the first long-range digital submarine cable in 1988. Prior to the invention of the fiber-optic cable, copper coaxial cables had been deployed in the Atlantic Ocean to connect the United States to Europe. The capabilities of these cables were very limited. Because of their limited capacity, they could hardly be said to have connected the DIs of individual nations to form a global DI. The first high-capacity fiber-optic submarine cable, TAT-8, connected the United States to Europe in 1988. It was built by AT&T with assistance from Bell Labs.[12] As with the other DI building blocks, the United States had an early technological lead, which it used to connect the U.S. NDI with the world. In contrast, China was not connected to other continents until a fiber-optic submarine cable linked it to Japan and the United States in 1999 (see Figure 4.1).[13] At the time, it was the longest submarine cable ever deployed.

Space

There are three categories of satellite networks: COMSAT, civil space-exploration, and military satellite-communications. It is easiest to discuss the first two types of satellites, which we focus on in this chapter. The development of space technology and satellites shows a clear U.S. lead—responsible for many early space developments—with China lagging behind significantly until the 21st century.

---

[11] Lionel Carter, Douglas Burnett, Stephen Drew, Graham Marle, Lonnie Hagadorn, Deborah Bartlett-McNeil, and Nigel Irvine N., *Submarine Cables and the Oceans: Connecting the World*, Cambridge, UK: United Nations Environment Programme, World Conservation Monitoring Centre, 2009.

[12] Tom Moylan, "Trans-Atlantic Telephone Cable Will Contain a Piece of L.V.," The Morning Call website, September 20, 1987.

[13] Paul Palumbo, "Undersea Fiber Network to Link China and U.S.," *Lightwave*, February 1, 1998.

Bell Labs built the first COMSAT and demonstrated it in orbit in 1962,[14] a feat China did not achieve until 1984, more than two decades later.[15] Similarly, the United States developed its first remote-sensing satellite, Corona, in 1960; China's came online in 1969.[16] In 1966, the United States introduced its first lunar orbiter, Lunar Orbiter 1, and achieved its first lunar soft landing, Surveyor 1.[17] These past space exploration missions and subsequent missions of the same kind have largely been remote controlled using extra-planetary satellite communications networks. China's first lunar orbiter was launched in 2007, with its first lunar soft landing in 2013.[18] The next three major space milestones came in 1971, 1976, and 1997 for the United States: first Mars orbiter (1971), first Mars landing (1976), and first Mars rover (1997).[19] China's first Mars orbiter, landing, and rover were not developed and launched until this year, 2021.[20]

U.S. export controls have been effective in preventing the Chinese acquisition of advanced satellite communications technology from the United States. At present, the U.S. DI in space (not U.S. owned, but U.S. made) is vastly bigger than China's. However, the Chinese robotic space program has progressed rapidly in the past decade, perhaps in part because of a success-

---

[14] Nokia Bell Labs, "Telstar 1," webpage, undated.

[15] NASA, "Chang'e 1," webpage, NASA Science: Solar System Exploration, April 5, 2019a.

[16] National Air and Space Museum, "Discoverer/Corona: First U.S. Reconnaissance Satellite," SpaceRace website, 2002.

[17] NASA, "Lunar Orbiter 1," webpage, NASA Science: Solar System Exploration, September 12, 2019f; NASA, "Surveyor 1," webpage, NASA Science: Solar System Exploration, August 11, 2019e.

[18] NASA, 2019a; Leonard David, "China Lands on the Moon: Historic Robotic Lunar Landing Includes 1st Chinese Rover," Space.com, December 14, 2013.

[19] NASA, "Mariner 9," webpage, NASA Science: Solar System Exploration, July 31, 2019d; NASA, "Viking 1," "Viking 1," webpage, NASA Science: Solar System Exploration, July 16, 2019b; NASA, "Mars Pathfinder," webpage, NASA Science: Solar System Exploration, July 25, 2019c.

[20] Mike Wall, "China's First Mars Mission, Tianwen-1, Successfully Enters Orbit Around Red Planet," Space.com, February 10, 2021; Planetary Society, "Tianwen-1 and Zhurong, China's Mars Orbiter and Rover," webpage, undated; Hanneke Weitering, "China's 1st Mars Rover 'Zhurong' Lands on the Red Planet," Space.com, May 14, 2021.

ful cyberattack campaign against NASA.[21] We discuss this in greater detail later.

Emergence of Competition

Internet

Capable telecom equipment suppliers, such as Alcatel, emerged in Europe in the 1990s. By the middle of the 1990s, the first internet connections had been established in China. The initial internet implementation in China was based on imported Western equipment. As mentioned earlier, the Chinese government developed policies to promote the development of an indigenous telecom industry that severely restricted Western suppliers' access to the Chinese telecom switching market and, later, to the market for internet switches and routers. The PRC permitted a small number of joint ventures between Chinese companies and Western suppliers. Initially, China formed a joint venture called Great Dragon between a consortium of state-owned companies and Siemens to produce telecom switches.[22]

Another state-supported firm was established in 1985 that also entered the market for telecom switches—ZTE. In contrast to the Great Dragon consortium, ZTE was privately managed. Another startup telecom firm, Huawei, was established a few years later, in 1987, and it too competed in the telecom switching market, allegedly sometimes bribing local officials to win contracts.[23] Eventually, both Huawei and ZTE competed effectively against the Great Dragon consortium and, by 2000, dominated the telecom switch market in China.[24]

---

[21] NASA, Office of the Inspector General, Office of Audits, *Cybersecurity Management and Oversight at the Jet Propulsion Laboratory*, Washington, D.C., IG-19-022, June 18, 2019.

[22] Eric Harwit, "Building China's Telecommunications Network: Industrial Policy and the Role of Chinese State-Owned, Foreign and Private Domestic Enterprises," *China Quarterly*, Vol. 190, June 2007.

[23] Harwit, 2007.

[24] Harwit, 2007.

Telecom

The first Chinese telecom carrier was formed in 1994, when two companies (the first Chinese telecom companies) spun off from the Chinese Ministry of Posts and Telecommunications as allegedly private companies independent of government control: China Telecom and China Unicom.[25] However, both probably received financing from Chinese state-owned banks, and it was not until 2002, nearly a decade later, that China Telecom was listed as a public company on the Hong Kong stock exchange.[26] Even then, the original China Telecom, formally known as the Chinese Telecommunications Corporation, retained a controlling interest in the new publicly listed China Telecom.

In the first decade or so, Chinese regulators made several attempts to break up the Chinese Telecommunications Corporation to reduce its monopoly power within China. The Chinese government came under pressure from the World Trade Organization (WTO) to comply with WTO rules and privatize China Telecom, a state-owned company. As a result, this pressure broke up China Telecom shortly before China entered the WTO in 2001, but the company remained state owned. Just as in the case of AT&T, several smaller companies were spun off to encourage competition; however, China Telecom retained dominant market share in the Chinese long-distance market, even after it was publicly listed in Hong Kong.

As a condition for joining the WTO, China agreed to let foreign companies take noncontrolling stakes in domestic telecom carriers in China, with foreign investors being permitted to hold up to 49 percent of a domestic telecom carrier.[27] To date, foreign investors have been unable to make significant investments in the major Chinese telecom carriers, although the

---

[25] The funding and control of many private Chinese companies are frequently opaque. Chinese accounting rules differ significantly from accounting best practices in the West and provide the opportunity for ownership interests to remain hidden, even for companies listed on stock exchanges in China, Hong Kong, and the United States. U.S. accountants have not been permitted to inspect the records of Chinese companies, that is, to conduct forensic accounting audits of Chinese companies.

[26] Goldman Sachs, "China Telecom Privatization Shines Through the Shadow of the Asian Financial Crisis," webpage, undated.

[27] "China Telecom History," Funding Universe website, Company Profiles, undated.

Chinese government has encouraged joint ventures to facilitate technology transfer from Western company suppliers to Chinese companies. According to a recent U.S. Senate subcommittee investigation, all three major Chinese telecom carriers—China Telecom, China Unicom, and China Mobile—are assessed to be government-owned enterprises.[28] China has retained a tight control over its domestic segment of DI.

The Chinese government also encouraged Chinese telecom carriers to expand globally.[29] In particular, China Telecom established subsidiaries in the United States in 2002, the UK in 2006, Germany in 2010, and France in 2013.[30] Chinese telecommunications now have over 200 foreign network PoPs outside China, in addition to an unspecified number of PoPs in the United States. Chinese carrier network PoPs in the United States provide direct connections from China to Chinese embassies and probably Chinese company facilities located in the United States. The PoPs may have been used in the hijacking of U.S. network traffic to China and may have supported cyber espionage campaigns conducted by Chinese hackers against the U.S. government and U.S. private-sector firms:[31]

> At least three other Chinese state-owned carriers have been operating in the United States for decades. The U.S. subsidiaries of the two other Big Three carriers—China Telecom and China Unicom—along with a smaller state-affiliated provider ComNet (USA) LLC ("ComNet") each received authorization to provide international telecommunications services in or prior to 2002 and have been operating ever since.[32]

Thus, evidence has emerged that the PRC has recognized the value of having a presence in the NDI of other countries, the United States in particular, to facilitate their own intelligence-gathering, espionage, and technology theft operations.

---

[28] U.S. Senate, 2020.

[29] U.S. Senate, Permanent Subcommittee on Investigations, 2020.

[30] China Telecom (Europe) Ltd., "History of China Telecom Europe," webpage, 2020.

[31] U.S. Senate, Permanent Subcommittee on Investigations, 2020.

[32] U.S. Senate, Permanent Subcommittee on Investigations, 2020.

The FCC is tasked with regulating U.S. telecom markets and the operations of both foreign and domestic telecom carriers. The U.S. Senate Permanent Subcommittee on Investigations has criticized the FCC for its lax oversight of Chinese telecom companies operating in the United States. Because of the national security risks identified with the operations of these carriers in the United States and the recognition that they are essentially state-owned enterprises, members of Congress have pressured the FCC to regulate these Chinese companies more carefully and to direct them to remove their network PoPs from U.S. territory.[33]

In contrast, U.S. telecom companies are not permitted to operate in China and have no PoPs in China. U.S. telecom carriers are also not as active in global telecom markets as Chinese carriers. At one point, AT&T did own or had significant stakes in several telecom carriers in Mexico but has since sold these stakes.

Application Layer

As mentioned in Chapter Two, the terrestrial network includes endpoints connected to the network, such as network control centers, data centers, cloud computing centers, and user devices. These network endpoints store data and enable users to create data using applications. Consistent with the Open Systems Interconnect model, we call this part of the terrestrial network the *application layer.*

The application layer of the terrestrial network has undergone significant changes and advances since the creation of the internet by U.S. research organizations and its later privatization, which did not fully occur until 2016. We illustrate the evolution of the application later in Figure 4.2.

Figure 4.2 shows that U.S. companies led in the development of key applications made available in the DI—email, the web browser, internet search, social media, and cloud computing. Initially, the U.S. research community played a significant role in the development of the internet. NSF created the NSFnet, which linked together research centers across the country. NSFnet contributed to the maturation of internet technologies and to early applications, such as the web browser.

---

[33] U.S. Senate, Permanent Subcommittee on Investigations, 2020.

FIGURE 4.2
Key Events in the Evolution of Application Layer

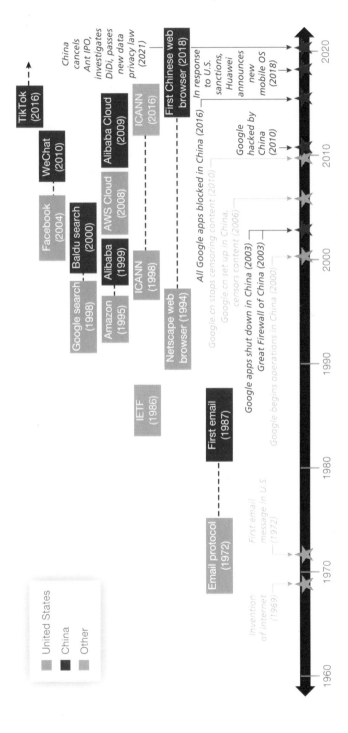

NOTES: AWS = Amazon Web Services; ICANN = Internet Corporation for Assigned Names and Numbers; IETF = Internet Engineering Task Force; IPO = initial public offering; OS = operating system.

In the early 1990s, commercial firms in the United States started building their own internet networks based on protocols matured in NSFnet. By 1998, some core internet functions were privatized, and NSF ended its direct role in the internet. In that year, the Department of Commerce formalized an agreement that gave a nonprofit, ICANN, control of the internet Domain Name System (DNS) and domain registration process. Initially, ICANN was under U.S. government oversight. ICANN was fully privatized and transitioned from U.S. government sponsored organization to an international one in 2016.

Figure 4.2 also shows the early history of the internet within China. In the first decade of the internet being available in China, applications were available only from U.S. companies, such as Google. However, in some cases, these applications enabled the Chinese people to gain access to uncensored information from Western media and other sources. This unfiltered access to information alarmed Chinese authorities and prompted them to shut down Google search in China and then to conduct a cyberattack against Google to find out whether Chinese dissidents were using Google email to organize against the Chinese government. The PRC blocked all Google applications in China in 2016. Since then, Google and many other Western technology companies have been unable to provide applications within China.

U.S. social media companies—Facebook, Twitter, and others—have never been approved to operate in China. Instead, as indicated in Figure 4.2, China was able to promote the development of its own social media companies, such as Tencent (WeChat) and Bytedance (TikTok). Use of these Chinese apps has spread throughout the world, affording these companies digital presence on many mobile phones and PCs worldwide. Only China has been able to foster the development of an internet application layer ecosystem with global reach, similar to the global reach and popularity of the application layer ecosystem created in U.S. companies.

U.S. technology giants emerged in the application layer: Facebook, Amazon, Apple, Netflix, and Google (FAANG), and some now include the older technology giant Microsoft in this group. The U.S. technology giants have benefited the most economically from the growth of internet services, including advertising and media services. Many of the U.S. technology giants now own DI, such as fiber-optic cables and other fiber-optic networks. RAND analysis indicates that the most successful U.S. technology firms have developed the "killer apps" of the application layer and have chosen to build the DI needed to support these applications. These DI companies and suppliers also potentially now have the greatest digital presence because of the data they have collected over the past several decades and the access they have to user devices and applications.[34]

The analogous Chinese technology giants include China Mobile, Tencent, Alibaba, Baidu, and Xiaomi. In both China and the United States, concerns have arisen over the vast stores of data that these technology giants have amassed in their data centers, not only of citizens in their own countries but of all their users throughout the world. U.S. data-privacy policy remains in flux and fragmented because of differing privacy laws in different states and the lack of an overarching federal privacy law. On the other hand, the PRC has just passed a new data-privacy law in an effort to curb an apparently growing black market for personal data in China and data theft and, possibly, to blacklist foreign data handlers and analysis firms that may be providing services to Chinese technology giants.[35] This new law will likely make it more difficult for foreign firms to gain a digital presence in China. However, the new law does not restrict the right of the Chinese government to access the data the Chinese technology giants collect.

Wireless

Wireless networks have advanced significantly from one generation to the next, now up to 5G networks. These network advances have been driven by private-sector telecom equipment suppliers and technology companies creating innovative mobile phones and other technologies for wireless networks. Early mobile phone innovators included Nokia of Finland and Blackberry of Canada. But their products fell victim to the smartphone. The first smartphone, the iPhone, was introduced by Apple CEO Steve Jobs in 2007. Today, the vast majority of mobile phones sold are smartphones. A Chi-

---

[34] Timothy M. Bonds, James Bonomo, Daniel Gonzales, C. Richard Neu, Samuel Absher, Edward Parker, Spencer Pfeifer, Jennifer Brookes, Julia Brackup, Jordan Willcox, David R. Frelinger, and Anita Szafran, *America's 5G Era: Gaining Competitive Advantages While Securing the Country and Its People*, Santa Monica, Calif.: RAND Corporation, PE-A435-1, 2021.

[35] Xiao, 2021.

nese company did not produce a smartphone until 2013. But now, in 2021, several Chinese companies, such as Oppo, Xiaomi, and Vivo, are all in the top-ten list of smartphone makers in terms of global market share. Chinese companies, including Huawei, have grown and become major players in the smartphone market.

As with wireline carriers, the PRC granted wireless network licenses only to domestic state-owned enterprises and, through regulatory barriers and other means, has not permitted foreign investment in these companies. In contrast, all U.S. wireless carriers are privately owned companies. In the past decade, several major U.S. wireless carriers are or were majority controlled by foreign investors. Before its merger with T-Mobile, Sprint was controlled by SoftBank of Japan. Even after its merger with Sprint, Deutsche Telecom has retained a significant ownership stake in T-Mobile, one of the largest wireless carriers in the United States.

Wireless networks require cellular infrastructure. Early on, important wireless infrastructure equipment suppliers emerged first in the United States and Europe, such as Motorola and AT&T in the United States and Alcatel, Ericsson, and Nokia in Europe.

The leading providers of cellular infrastructure have changed over time and, at one time, included a number of U.S. companies, such as Motorola and AT&T. Lucent was originally a division of AT&T but was spun off in 1996 and, at first, flourished as an independent company, as telecom carriers raced to add wireline capacity to increase internet bandwidth and wireless capacity and coverage to their new networks. Lucent, however, did not keep up with technology advances in the internet routing and switching market. The company lost market share and suffered financial losses in the so-called dot-com bust at the turn of the century and was ultimately sold to Alcatel.[36] The merged company was called Alcatel-Lucent, which was headquartered in Europe, then was sold to Nokia. For a time, Alcatel-Lucent was a major supplier of wireless infrastructure equipment. After 2010, when Motorola's wireless infrastructure division was sold to Nokia, the United States no longer had any domestic wireless infrastructure vendors (see Figure 4.1). Nortel of Canada was another leading terrestrial and wireless

---

[36] "Alcatel to Acquire Lucent in $13.4 Billion Deal," MSNBC, April 3, 2006.

infrastructure supplier, providing base stations and other equipment to U.S. wireless carriers. Nortel ultimately filed for bankruptcy protection in 2009 and ceased operation.

Later reports surfaced that Nortel had suffered a major cybersecurity breach.[37] Investigators found that Chinese hackers had had access to Nortel networks and servers for more than a decade and had stolen critical intellectual property and trade secrets.[38]

The Nortel hack was not the only purported cyber incident affecting Western telecom suppliers. In 2002, Cisco Systems sued Huawei for the alleged theft of software from Cisco's corporate network. Cisco showed in court that the software in Huawei routers contained code that closely resembled the code in Cisco routers and that the Huawei code contained the same bugs present in the Cisco software. The U.S. court case between Cisco and Huawei was settled in 2004.

It was during the late 1990s and early 2000s, when U.S. and Canadian telecom suppliers suffered cyber breaches, that Huawei rose to prominence and became a major global supplier of first wireline and later wireless telecom equipment. Huawei's share of the wireless infrastructure market grew rapidly during this time. By some accounts, Huawei is now the largest telecom equipment supplier in the world. In 2019, it reported revenues of $17 billion and operated in 170 countries throughout Africa, the Middle East, Europe, and Asia.[39]

Huawei was able to increase its market share by consistently underbidding European and Asian competitors, sometimes by up to 40 percent. It has been reported that Huawei has received over $70 billion in various subsidies from the Chinese government over the past few decades to support its busi-

---

[37] Natalie Obiko Pearson, "Did a Chinese Hack Kill Canada's Greatest Tech Company?" *Bloomberg Businessweek*, July 1, 2020.

[38] Ryan Naraine, "Nortel Hacking Attack Went Unnoticed for Almost 10 Years," ZDNet, February 14, 2012.

[39] Keith Johnson and Elias Groll, "The Improbable Rise of Huawei," *Foreign Policy*, April 3, 2019.

ness operations.[40] These subsidies helped finance the purchase of Huawei equipment by wireless carriers, especially in developing countries.

The cellular infrastructure market has undergone further consolidation in the past decade. In 2015, Nokia acquired Alcatel-Lucent to better compete with Ericsson and its Chinese rivals, Huawei and ZTE.[41] Meanwhile, Huawei has maintained its leading position as a supplier of 4G and 5G infrastructure equipment.

For decades, the U.S. government has let free-market forces operate unencumbered in both the wireless carrier and cellular equipment markets. This stance appears to have served the country well in mobile phones; U.S. companies have maintained a significant presence in markets for mobile phones, mobile software and applications, and the design of leading-edge microchips for 5G phones. However, as noted earlier, the United States no longer has a prime systems integrator in the wireless infrastructure market. The United States has to rely on European manufacturers (Ericsson and Nokia) and a Korean company, Samsung, to supply trusted cellular infrastructure equipment for networks. In contrast, the evidence we have presented suggests that the PRC views both terrestrial and wireless telecom markets as strategic economic markets with national security implications. This view is supported by the following reported exchange between Ren Zhengfei, founder and CEO of Huawei, with Jiang Zemin, then–General Secretary of the CCP:

> I said that switching equipment technology was related to national security, and that a nation that did not have its own switching equipment was like one that lacked its own military. Secretary Jiang replied: Well said.[42]

---

[40] Chuin-Wei Yap, "State Support Helped Fuel Huawei's Global Rise," *Wall Street Journal*, December 25, 2019.

[41] Griffith, 2015.

[42] Harwit, 2007.

Submarine Cables

The first fiber-optic submarine cables were financed by governments or by major telecom carriers. Chinese telecom companies, which are state owned, have financed and co-own a growing number of submarine cables connecting China to other countries in the Pacific region and elsewhere in the world. PRC planning documents for its DSR plan identify the objective of expanding the network of submarine cables in the oceans of the world. Moreover, in its Made in China 2025 plan, the Chinese government set a goal for Chinese telecommunications and industry to capture 60 percent of the global fiber-optic market by 2025.[43] While U.S. telecom companies have retreated from the submarine cable market, reportedly because of the sizeable capital expenditures required, Chinese telecom firms have increased their ownership stakes in the submarine cable market. Fortunately, U.S. technology giants, such as Google, Microsoft, Amazon, and Facebook, have continued to finance the development and deployment of cable systems, although some Western observers predict that China telecommunications are poised to dominate the global market.[44]

The three largest companies that can build and deploy submarine cables are Alcatel Submarine Networks of Alcatel-Lucent (France); TE SubCom (Switzerland); and NEC Corporation (Japan).[45] By some estimates, Huawei Marine is the fourth-largest provider of submarine cable systems and is very active in Africa. Huawei Marine has been accused of receiving subsidies from the PRC and providing submarine cable systems below cost to customers.[46] Although U.S. companies were early leaders in the development of submarine cable technologies and although AT&T's Lucent division helped deploy the first fiber-optic submarine cable in 1988, U.S. companies are no longer major players in the submarine cable supplier or installation markets.

---

[43] Francesco Bechis, "Undersea Cables: The Great Data Race Beneath the Oceans," webpage, Italian Institute for International Political Studies (ISPI), May 31, 2021.

[44] Winston Qiu, "Why It Is China's Turn to Lead the Submarine Cable Industry," Telecom Ramblings website, February 11, 2014.

[45] Brake, 2019, p. 11.

[46] Brake, 2019.

In the Pacific region, most submarine cables have joint ownership, in which companies from different countries jointly own a single cable, although a single company may operate the cable system at all landing points. Most submarine cables landing in the United States are co-owned by U.S. companies. Similarly, most submarine cables landing in China are co-owned by Chinese companies, in particular, state-owned Chinese telecom companies.

Space

For many years, the United States had a substantial lead over China in COMSATs. There has been very limited transfer of satellite communication technology from the United States to China because much of this technology is export controlled. This advantage in commercial satellites has enabled the United States to remain a major supplier of such satellites in the world market. In addition, because the major COMSAT constellations are owned by U.S. allies, the U.S. government and U.S. military have access to a wide range of COMSATs, including those owned by Intelsat and Inmarsat, in addition to DoD's military COMSATs. In 2019, some experts estimate that U.S. companies and the U.S. military have access to over 580 Gbps of satellite communication bandwidth, whereas China had access to only 20 Gbps from its own COMSATs.[47] This predicament has forced the PRC to use U.S.-made COMSATs to support its military forces deployed to the South China Sea.[48]

In other areas of space technology, the story appears different. What enabled China to suddenly catch up to the United States and space exploration technology? One hypothesis suggests that China has been able to acquire intellectual property and export-controlled technology from NASA by means of cyberattacks, such as those it allegedly executed against Cisco Systems in Nortel, which we described earlier in this chapter. Indeed, significant evidence indicates that China successfully penetrated the networks of NASA and its key R&D organization, the Jet Propulsion Laboratory.[49] These

[47] Spegele and O'Keeffe, 2019.

[48] Spegele and O'Keeffe, 2019.

[49] NASA Office of the Inspector General, Office of Audits, 2019.

cyberattacks took place for almost a decade, from 2009 to 2018. It therefore appears that the PRC had long-term and extensive access to at least some internal networks and servers where vital NASA information was stored. It has also been reported that Chinese operatives have attempted to acquire satellite technology from Boeing.[50] In addition, the DoD Inspector General has recently reported that some Air Force space programs were vulnerable to cyberattack.[51] If China has been exploiting the cyber vulnerabilities of U.S. commercial, civil, and military space programs, the information and technology it has gained may have enabled the significant advances in Chinese civilian space activities observed in 2021 (i.e., a successful soft landing of a robotic spacecraft and rover on Mars) and may enable China to close the gap in the satellite communications segment of DI in the coming decade.

Foundational Elements

Microchips

As was the case with DI network segments, the United States was an early technology leader in development of the foundational elements for DI. The first microchip was invented in the United States in 1958 (Texas Instruments),[52] as were the first semiconductor microchip (Fairchild Semiconductor)[53] and the first microprocessor with a programmable instruction set (Intel).[54] The United States quickly became a leader in both the design and manufacture of microchips.[55]

---

[50] Brian Spegele and Kate O'Keeffe, "China Maneuvers to Snag Top-Secret Boeing Satellite Technology," *Wall Street Journal*, December 4, 2018.

[51] Sandra Erwin, "Defense Inspector General Finds Key Air Force Space Programs Vulnerable to Cyber Attacks, Sabotage," webpage, SpaceNews, August 16, 2018.

[52] Royal Irish Academy, "Evolution of the Microchip," webpage, July 27, 2020.

[53] Royal Irish Academy, 2020.

[54] Intel, "Intel at 50: The 8086 and Operation Crush," press release, May 31, 2018.

[55] Microchips are manufactured using sophisticated semiconductor fabrication techniques. Billions of interconnected transistors are printed onto a semiconductor wafer to form a microchip. One example of a microchip is the microprocessor or central process-

Over time, however, microchip manufacturing moved overseas, pre-dominantly to Asia. Two important companies were founded in Asia: Samsung Electronics, in 1980,[56] and TSMC, in 1987.[57] Today, these have become the undisputed leaders in state-of-the-art microchip manufacturing. As fewer microchips were being manufactured in the United States, DoD became concerned about ensuring it had a secure source for trusted and unaltered microchips for use in U.S. weapon systems. The first DoD program to address this started in 1990, when the National Security Agency established its own state-of-the-art microchip foundry.[58] However, because the cost of building and maintaining the foundry was so high, DoD decided to close it in 2004. The DoD needs relatively few microchips compared with the massive numbers that go into commercial and consumer products.

DoD established the Trusted Foundry program in 2005, which began with a single foundry located in Fishkill, New York, owned by IBM. For almost a decade, this arrangement appeared to satisfy DoD's microchip needs. During this time, a number of other trusted foundries were also established in the United States. IBM, however, was reportedly losing a substantial amount of money each year at the Fishkill foundry and, in 2014, decided to sell the foundry to a newly formed company called Global Foundries, which is owned by foreign investors from the United Arab Emirates. This investor group received clearance to purchase the foundry after agreeing to certain safeguards. With IBM's assistance, Global Foundries remained at or near the state of the art in microchip manufacturing until 2018, when it suspended development of its 7-nm process node. This was effectively an acknowledgement that it had fallen seriously behind Samsung and TSMC and was unlikely to catch up or did not have the capital to continue developing this process. The DoD Trusted Foundry program has remained at

---

ing unit of a computer. Another is a microcontroller, which can be used to control some systems on a satellite or in an automobile.

[56] Geoffrey Cain, *Samsung Rising: The Inside Story of the South Korean Giant That Set Out to Beat Apple and Conquer Tech*, New York: Penguin Random House, 2020.

[57] TSMC, "Company Info," webpage, undated.

[58] Advanced microchips are manufactured in facilities called *foundries* using a series of manufacturing processes and specialized equipment and materials.

14 nm. Meanwhile in 2017, both TSMC and Samsung moved to 5-nm process nodes and in 2021 are producing microchips at 5 nm.[59]

The United States has one other major semiconductor fabrication company of note, Intel. Until recently, Intel was able to stay on par with the industry leaders in Asia. However, it encountered serious difficulties in maturing its 10-nm process node. Intel's 10-nm microchips were delayed several years as it worked to improve chip yields to acceptable levels. Its 10-nm microprocessors become widely available only in 2021. Then, in 2021, Intel announced it would outsource some of its microchip production to TSMC, the first time in its history that Intel has outsourced chips. The United States thus no longer has state-of-the-art microchip foundries on its own soil and relies on third-party foundries in Asia for production and final fabrication of the most advanced microchips.

China has made it a strategic economic and national security objective to develop its own indigenous microchip designs and fabrication industries. China's most important semiconductor company, the Semiconductor Manufacturing International Corporation, was founded in 2000 and can produce chips at 14 nm.[60] Despite pouring billions of dollars into microchip development, China remains well behind both the United States and the other state-of-the-art industry players. While U.S. firms lagged two generations behind the industry leaders, TSMC and Samsung, as of late 2021, Chinese firms lagged even further behind.

It should be noted, however, that the United States retains a significant lead technologically in the design of advanced microchips and design tools and is still a major supplier of the microchip fabrication equipment that both Samsung and TSMC rely on to build their most advanced chips. The export of state-of the-art U.S. and European microchip fabrication equipment has been restricted and is not available to Chinese companies, a policy that has

---

[59] The entire manufacturing process with the associated equipment is termed a *process node* in the semiconductor industry. Although associating a single technical feature with a particular process node is an oversimplification, a node is commonly associated with the smallest line with that the manufacturing process can produce. The most advanced process nodes available commercially in 2021 have minimum lines of 5 nm.

[60] Semiconductor Manufacturing International Corporation, "Overview," webpage, undated.

hampered the development of the Chinese microchip industry. Western policymakers fear that, if China were one day to dominate the microchip industry, it could effectively control each of the DI building blocks, as well as the computer devices and mobile phones that connect to DI, which could have significant and pervasive economic and national security implications.

Technical Standards

Technical standards underpin the communication systems that make up DI. These technical standards define how equipment from different vendors should interconnect to establish communication links, how network resources are managed, and how other functions should be provisioned. Different technical standards are used in each DI building block or network segment. International standards setting bodies, such as the 3rd Generation Partnership Project (3GPP),[61] set technical standards.[62] Technical experts from companies supplying or operating DI networks that compete for market share against one another work together in such organizations as 3GPP to develop technical standards. Consequently, as U.S. companies lose market share or exit DI infrastructure supplier markets altogether, the United States may have less insight into and influence over emerging technical standards of new DI networks, such as 5G.

Typically, companies, not countries, develop technical standards. In the 1990s, U.S. companies played a larger role than Chinese companies did in standard development because of the technology leadership of U.S. companies. Technical standards are usually agreed on by consensus or vote in most standards-setting organizations. Technology and patents constitute important ingredients for standards. A system essential patent (SEP) for a cellular network defines a core function of the network that should be used to implement and connect to such a network. Companies try to include their technologies or intellectual property in SEPs and will therefore offer their technology and intellectual property to define a candidate technical standard. SEPs may represent valuable intellectual property and may require equip-

---

[61] 3GPP is the international standard-setting organization that is responsible for developing and approving the technical standards for cellular communication networks.

[62] In satellite networks, proprietary standards are still sometimes used.

ment vendors to purchase subsystems or microchips from the SEP holder if the vendors cannot reproduce the functionality of the subsystem using their own microchips or other components. However, the subsystems or microchips may contain additional functionality and undocumented capabilities. An untrustworthy SEP holder can exploit a network or, possibly, take control of it using undocumented subsystems in the holder's products. SEPs and technical standards related to the security architecture of DI networks raise special concern.

In the past, such concerns did not arise with 3G or 4G cellular networks. However, in the past decade, Chinese companies have played a much larger role in developing technical standards for 5G networks through participation in 3GPP. In addition, China now plays a bigger role in developing intellectual property pertaining to advanced networking technologies and, by one estimate, led the world in patent applications in 2019, narrowly beating out the United States in this category.[63] However, the quality of these patents has been questioned. The majority of Chinese patent applications, at least in some fields, are filed only in China, not in Western countries. These domestically filed Chinese patents may not be approved in Western countries. However, these domestic patents can potentially be used to shield Chinese companies from foreign competitors, even ones that possess superior DI-related technologies, thereby preventing Western DI systems and components from permeating the Chinese NDI.

Given some measure of control or ownership over these foundational elements, China could render them untrustworthy to the United States or could deny the United States access to these technologies. Microchips and complex technical standards lie at the heart of all elements of DI, and trustworthiness of these foundational elements will therefore have to be monitored.

---

[63] Miguel Bibe, "China Takes Over World Leadership in International Patent Applications," webpage, Inventa International, April 22, 2020.

Trends and Asymmetries

Digital Infrastructure Building Blocks

Cellular

Chinese companies are significant players in the smartphone market; however, Apple and Google provide trusted mobile devices and have significant market presence. The United States maintains access to trusted mobile devices, while the majority of PRC users use Android phones. For wireless infrastructure, Huawei has a growing market presence, whereas U.S. presence in this infrastructure market ended in 2010. Chinese firms have the potential to dominate the market as Huawei, ZTE, and customers continue to receive favorable financing terms for new projects and, perhaps, other state subsidies. On the other hand, Nokia, a European company, has struggled in developing competitive 5G products and has, at times, encountered financial losses.

International

Both China and the United States have significant digital presence in recent and planned Pacific submarine cables. Additionally, China is extending its reach and access by having its telecom companies establish network PoPs globally. The U.S. telecom companies have a more domestic focus, which changes the nature of certain intelligence activities.

Terrestrial

The United States dominated in the early phases of DI; however, China has expanded its presence in computers, servers (e.g., Lenovo), routers, and switches (e.g., Huawei). The United States still has access to state-of-the-art trusted terrestrial technologies (e.g., Cisco, Intel, Hewlett Packard, Dell, Microsoft). China has its own trusted suppliers but remains reliant on Microsoft for computer operating systems for consumer PCs.

Space

Foreign companies, under the control of allied nations, own the majority of COMSATs on orbit. In addition, advanced U.S.-owned COMSATs are emerging (e.g., SpaceX Starlink). Furthermore, the United States remains a technology leader with access to advanced COMSATs. The PRC remains focused on deploying military satellites and antisatellite capabilities.

We now examine a few structural trends and asymmetries that emerged from DI's evolution.

Foundational Elements

PRC and U.S. trends in DI foundational elements (microchips, standards) and applications (i.e., TikTok, Google) reflect overall trends in the U.S.-PRC DI competition.

Microchips

Until recently, the United States was the world leader in manufacturing microchips and remains a world leader in chip design. Today, the most valuable U.S. chip companies, Nvidia and Qualcomm, do not have their own foundries. However, the United States, including its most valuable DI-related companies, now relies on TSMC and Samsung for the manufacturing of state-of-the-art chips. China has designated microchips as a key technology, investing billions to become self-sufficient in not only the design but also the manufacturing of microchips. However, several recent expert estimates suggest the PRC lags at least a decade behind leading (non-U.S.) companies.

The U.S. government has recently recognized the importance of establishing state-of-the-art microchip foundries in the United States. The U.S. government struck a deal with TSMC to build a chip foundry in Arizona. Samsung is considering building a foundry in Texas capable of making not just memory chips but also the most advanced logic chips needed for 5G DI. The U.S. Senate and the Biden administration have proposed creating a new program that will provide $50 billion in incentives to encourage Western chip makers to build state-of-the-art foundries in the United States. However, these proposals have not become law (as of August 2021). Intel, which has recently struggled to produce microchips of the same caliber as those produced by TSMC and Samsung, has proposed that the U.S. government help reinvigorate its microchip R&D program and build a new foundry in Arizona. At present, Intel's chip manufacturing capability may be three years behind TSMC.

Technical Standards

During the early stages of DI development, private industry in the United States and Europe led the development of DI standards; however, for a variety of reasons, DI standards that are set by international organizations and companies headquartered in many countries, including China, now play important roles in setting DI technical standards. As U.S. technology leadership has eroded and as U.S. companies have exited certain markets, such as that for cellular infrastructure, it has become more difficult for U.S. companies to lead development of important DI standards.

Application

China tightly controls its application layer, with many U.S. companies being banned from operating in the PRC even though the majority of PRC companies can operate in the U.S. application layer. After the internet was established, U.S. companies dominated application development and reaped the economic benefits (e.g., business applications, global e-commerce, social media and search markets). After China denied U.S. tech giants from operating in China, the PRC was able to foster the development of its own tech giants, many of which now have global reach (e.g., TikTok).

Structural

The evolution of DI also demonstrates that a critical way to create advantage in DI comes from the relationship between a country's public and private sectors. Washington and Beijing have distinct and disparate approaches to the public-private relationships. The CCP has developed initiatives that both compel cooperation and subsidize PRC companies to create national DI champions (e.g., Huawei). On the other hand, the U.S. government's relationship with the private sector has shifted over time.

In the early phases of DI, U.S. government R&D enabled the government to gain insight into DI and to promote U.S. technology and market leaders. After WWII, the U.S. government dedicated billions for R&D to build scientific infrastructure and advanced dual-use technologies, such as microchips. However, spending as a proportion of gross domestic product has declined over time. In 1964, the U.S. government spent 1.86 percent of gross domestic product on R&D; by 1994, it was spending 0.83 percent; and in 2019 it spent 0.6 percent.[64] This demonstrates a gradual shift downward in terms of U.S. government R&D spending, represented as a proportion of gross domestic product. Furthermore, U.S. tech companies, investors, and the U.S. government have deemphasized important but capital-intensive industries, such as microchip manufacturing, over time. Until recently, the U.S. government has provided few incentives to support the construction of foundries in the United States.

In the 1970s, venture capital firms emerged in the United States; by the 1990s, venture capital–style investment firms had also emerged, allocating capital that was not necessarily tied to U.S. interests or U.S. national-security needs.[65] This trend—growing differences between U.S. government and private-sector priorities for technology innovation and sharing—continued into the 21st century. Because of this trend, despite recent efforts to improve information-sharing between the U.S. government and Silicon Valley, the U.S. government may not be aware of all of the R&D of the U.S. tech giants (e.g., FAANG and other companies, such as Qualcomm).[66] In contrast, the CCP has directed, cultivated, and leveraged PRC companies for military, economic, and domestic purposes.

Summary

We conclude this chapter with a summary technical comparison of the relative U.S. and PRC positions in DI and DI foundational elements described in the previous section. We compare two periods, the mid-1990s and early 2020s.

---

[64] Christopher Darby and Sarah Sewall, "The Innovation Wars: America's Eroding Technological Advantage," *Foreign Affairs*, March–April 2021.

[65] Darby and Sewall, 2021.

[66] Some recent initiatives of this kind include the creation of DIUx and In-Q-tel by DoD and the U.S. government.

Historical: Mid-1990s

The United States was an early leader in developing DI building blocks and foundational elements in the mid-1990s. The United States built many initial versions of DI, and our historical assessment reflects this:

- **International.** U.S. companies first developed a digital fiber-optic submarine cable in 1988, which became a key enabler for communication. Conversely, China did not develop its first digital fiber-optic cable until the late 1990s. The United States led in developing and operating digital submarine cables, with greater presence and leadership giving the country an early advantage. For PoPs, U.S. companies' network PoPs emerged in the mid-1980s but were limited overseas, with none in China. China's PoPs emerged slightly later, in the mid-1990s, and China established a larger overseas presence over time.
- **Terrestrial.** The United States led the development of DI in the terrestrial layer. U.S. companies were early leaders and suppliers of terrestrial infrastructure globally, with PRC companies in the terrestrial layer not emerging in global markets until the early 1990s. The 1990s illustrate an early U.S. lead in terrestrial networks and infrastructure, given its role as an initial technological leader and supplier globally.
- **Cellular.** U.S. companies were the initial providers of cellular network infrastructure. The PRC did not emerge in the cellular layer until 1996, when Huawei introduced the first PRC wireless network. Although this trend had shifted by the 2020s, the United States had an early lead in cellular infrastructure and networks.
- **Space.** U.S. companies were the early developers and suppliers of COMSATs and other space technology. The first PRC COMSAT was launched in 1984, but China trailed the United States in COMSATs and satellite technology. Similarly, the PRC did not match the United States in robotic space advances until 2021. The United States had a robust robotic space program by the 1970s.
- **Foundational elements.** For microchips, DoD was the driver for semiconductor R&D from the 1960s through the 1980s, as well as for the first consumer computers. The flagship PRC microchip firm, SMIC, was not founded until 2000. Furthermore, the United States was an

early world leader in chip designs and fabrication tools, with PRC firms being reliant on U.S. chip designs and fabrication tools.

When translating these high-level technical comparisons into relative OAC of DI, a picture emerges depicting strong U.S. advantage. OAC alone does not necessarily confer advantage, but we argue that the extent of OAC in DI does offer insight into a country's potential to develop and use digital presence in another country's NDI while simultaneously retaining the integrity of its own. Table 4.1 provides a snapshot of the relative degrees of OAC of DI that the United States and China held in the mid-1990s. We note that this provides a preliminary analysis and does not intend to be comprehensive of all factors that dictate relative advantage. Table 4.1 focuses solely on technical aspects of OAC and DI and excludes structural factors.

Current: 2020s

In the past decade, China has strengthened its position in DI building blocks, except for space. For foundational elements, China continues to lag in microchips and in technical standards and patents. While it has reduced the positional advantages of the United States in the latter category, China

TABLE 4.1
Historical Ownership, Access, and Control Technical Comparison, 1990s

	United States	China
DI building blocks		
International	Advantage	Mixed
	Slight advantage	Slight disadvantage
Terrestrial	Advantage	Slight disadvantage
Cellular	Slight advantage	Disadvantage
Space	Advantage	Disadvantage
Foundational elements		
Microchips	Advantage	Mixed
Other foundational	Advantage	Disadvantage

continues to be in a position of relative disadvantage vis-à-vis the United States in perhaps the most important foundational element—microchips:

- **International.** Each country has equal positions in submarine cables. U.S. companies own a significant array of cables, and U.S. landing sites provide access to message traffic. PRC ownership of cables in the U.S. Indo-Pacific Command area of responsibility is increasing. PRC landing sites provide access to message traffic. However, PoPs are a different story, emerging over the past 30 years. U.S. companies continue to have a small number of network PoPs overseas and none in China. In contrast, the PRC has hundreds of PoPs globally, including in U.S. telecommunications networks.
- **Terrestrial.** Each country has equal positions in terrestrial network segments. Both U.S. and Chinese companies are major suppliers of terrestrial infrastructure globally.
- **Cellular.** We assess the United States and China to have relative parity in the cellular network segment of DI (this assessment applies globally). U.S. policy protects U.S. NDI from untrustworthy Chinese DI equipment, putting the United States in a position of advantage. The United States has very limited presence in cellular networks globally because it does not have infrastructure suppliers in the global market; China has a significant global market presence. However, U.S. restrictions on Huawei access to advanced 5G chips have hampered the company's supply chain for 5G infrastructure systems to some degree. The United States also still has access to trusted suppliers, such as Ericsson, Nokia, and Samsung.
- **Space.** The United States retains a position of advantage in space. China is attempting to close the gap but has had limited success so far when it comes to satellite networks. U.S. companies are major suppliers of COMSATs and lead development of advanced COMSATs. Additionally, the United States has access to many allied systems. On the other hand, China still trails the United States in COMSATs and satellite technology.
- **Foundational elements.** The DoD and U.S. commercial firms continue to rely on Asian foundries for state-of-the-art microchips (7 nm). PRC firms have restricted access to Asian foundries for state-of-the-art

microchips (7 nm), and some firms, such as Huawei, have been black-listed. Furthermore, the United States still leads globally in advanced chip designs and fabrication tools. As a result, PRC firms rely on U.S. chip designs and fabrication tools. PRC firms also have an increasing portfolio of patents relating to DI standards. U.S. companies still possess a large portfolio of patents, some of which relate to international DI standards.

Table 4.2 provides a snapshot of how current technical comparisons between the United States and China in DI translate into relative degrees of DI OAC. The table shows that, although the United States does not have the same clear advantage it did in the 1990s, that does not mean it finds itself disadvantaged vis-à-vis China today in terms of DI OAC. Additionally, some of the changes from the 1990s to today—a less dominant U.S. position within DI—reflect typical progressions in technology; leadership often becomes more diffuse over time, with fewer sole leaders.

TABLE 4.2
Current Ownership, Access, and Control Technical Comparisons

	United States	China
DI building blocks		
International	Advantage	Advantage
	Slight advantage	Advantage
Terrestrial	Slight advantage	Advantage
Cellular	Parity	Parity
Space	Advantage	Disadvantage
Foundational elements		
Microchips	Slight advantage	Mixed
Other foundational	Slight advantage	Mixed

CHAPTER FIVE

Findings, Implications, and Emerging Opportunities

In this report, we identified and defined DI, its foundational elements, and the concept of digital presence and explained why they matter for the United States and China. We characterized the ongoing competition for DI and outlined key assumptions tied to China's pursuit of the ability to project power globally, the U.S. military's reliance on DI, and the risks and rewards associated with control of and access to DI building blocks and foundational elements. Our analysis yielded significant implications and potential opportunities for DoD. Consistent with traditional DoD net assessments, we intended to be diagnostic, not prescriptive; we have therefore focused our findings on characterizing the issue and diagnosing the implications for DoD. In this preliminary research effort, we do not propose recommendations or solutions.

Digital Infrastructure Will Play a Significant Role in Shaping the U.S.-China Competition

While traditional elements of power (military, economic, information, diplomatic) shape competition and conflict, power centers within each element are shifting toward DI:

- **Military.** Microchips underpin advanced weapon systems and operational concepts and affect overseas posture.
- **Economic.** DI companies and technologies represent a greater share of economic activity.

- **Information.** DI underpins the transmission, exploitation, and use of information for intelligence operations.
- **Diplomatic.** DI underpins many diplomatic efforts to build bilateral and multilateral relationships and plays a greater role in international institutions.

As a result of these shifts, control of and access to DI offer potential for the United States or China to create competitive advantages—in some instances, asymmetric advantages—vis-à-vis each other. Although all dimensions of power play a role in shaping competition, when it pertains to DI, we observe China and the United States competing strategically primarily in the economic and diplomatic realms (e.g., B3W versus BRI). Such diplomatic and economic initiatives related to DI have national security implications because of the potential for digital presence and the dual-use nature of DI.

While the United States Recognizes Digital Infrastructure as Driving Economic Growth, the U.S. Political-Economic Model May Not Be Poised to Compete with China's Approach

China and the United States recognize DI building blocks and foundational elements (chips, standards) as important to economic growth; however, the traditional U.S. political-economic model of a regulatory state and a laissez-faire form of capitalism may not be poised to compete effectively against the PRC in DI. China uses a statist economic model focused on centrally directed infrastructure investment initiatives. Through the BRI and DSR, for example, China attempts to build the NDI of other countries using Chinese products. This approach offers an opportunity for the PRC to elevate Chinese technological leadership, information-collection capabilities, and global influence. DI remains a focal point of China's multifaceted model for economic growth and enables Chinese companies to gain sizeable market share in DI-related sectors. The Chinese government uses an industrial policy and government subsidies for state-owned enterprises and private companies to build DI domestically, which the PRC extends globally through BRI and DSR initiatives. In contrast, after initial government R&D support, the U.S. private sector developed early iterations of DI with minimal government involvement.

U.S.-Led International Order Offers Structural Advantages to Shape the Competition for Digital Infrastructure

Long-standing, trusted U.S. partnerships and alliances and leadership of international institutions provide key structural advantages that may shape the competition in favorable ways for the United States. The United States exerts considerable influence over the international order that, we hypothesize, China potentially seeks to lead.[1] Many of China's cross-cutting strategic efforts identify development partners that rely on China for dual-use technologies and infrastructure through such initiatives as BRI and DSR and that are often funded through the China Development Bank. China's approach to diplomacy appears DI- and infrastructure-focused. On the other hand, the United States uses a more multinational approach predicated on long-standing international institutions, such as the International Monetary Fund, World Bank, WTO, United Nations, and the North Atlantic Treaty Organization.

The United States has already begun to leverage these partnerships and institutions as mechanisms for DI competition with China. For example, the B3W initiative leverages G7 countries to provide an alternative to Chinese infrastructure efforts. The enduring legitimacy of these institutions and their ability to offer attractive alternatives to China's model are essential for maintaining this advantage.

---

[1] Chinese leadership would likely be different from the U.S. leadership that has been in place since post-WWII and could require some revision of the order to align with Chinese interests.

Digital Infrastructure Ownership, Access, and Control by an Untrustworthy Actor May Introduce Risk into a Country's National Digital Infrastructure

A country or region with a DI dominated by an untrustworthy actor may present unacceptable risks to some military operations or governmental activities. We see the potential for this to occur in areas where Chinese companies have fielded DI.

DI equipment in other countries can provide an enduring and pervasive digital presence used for activities, such as intelligence collection. For example, Chinese OAC of DI in U.S. coalition partners or allies presents intelligence risks to U.S. companies, the U.S. military, and U.S. diplomats operating in those countries. While some aspects of these risks may be known to at least the U.S. government, the risks multiply and may be harder to counter if the PRC were to broaden and deepen its DI OAC.

Although an allied country may be aware of the risks posed by untrustworthy DI components supplied by a PRC company, if PRC companies were to one day dominate markets for specific types of DI equipment (e.g., 5G), then that country may have no choice but to introduce untrustworthy components into its NDI. Currently, the United States and other countries have access to trustworthy 5G equipment from several vendors. However, these vendors have at times suffered financial losses, putting into question their long-term health. It will be important for the U.S. government and allied governments to track the health of trustworthy suppliers in their DI supply chains to prevent dominance of these supply chains by an untrustworthy actor.

How Digital Infrastructure Evolves May Affect Warfare in Substantial Ways

DI and foundational elements have already advanced military capabilities, with future iterations of DI having the potential to affect warfare in substantial ways. The United States and China have efforts underway to leverage DI and foundational elements for military applications. Consistent with how military innovations and RMAs emerge, DI could potentially yield significant military opportunities for major military powers. Because of changes in military systems, operational concepts, and some organizational restructuring tied to DI, we expect DI building blocks and foundational elements to affect the character and conduct of warfare. As a result, OAC of DI will become increasingly important for military effectiveness.

Characteristics of the Future Security Environment Are Shifting Due to Digital Infrastructure

Characteristics of the future security environment will likely shift because of changes in OAC of DI and its foundational elements and because of changes in the digital presence of adversaries. These changes may have an effect on future platforms and capabilities. Key shifts in the future security environment may include the following:

- A country's physical footprint may no longer be sufficient to offer key military advantages in some scenarios.
- There is more than one way to project power; previous assumptions associated with power projection are being challenged.
- There is a growing rise and importance of dual-use technology and capabilities.
- Democratization of signals intelligence and the ubiquity of open-source intelligence
- Proliferation and reduction in size of precision strike weapons because of the availability of high-performance chips
- New EW and cyber capabilities that will render C3 networks more vulnerable.

Digital Infrastructure Plays a Distinct Role in U.S. and Chinese Visions of Power Projection

The U.S. approach to power projection, predicated on post-WWII–era assumptions and reliant on traditional military capabilities, differs from China's likely use of traditional and nontraditional means to project power beyond its borders. Furthermore, the roles of DI for the U.S. and PRC visions of power projection also seem to differ.

DI and digital presence appear central to PRC power-projection ambitions and potentially challenge traditional U.S. approaches to power projection. China's traditional military power-projection capabilities have grown in recent years but remain limited. However, DI and digital presence may provide the PLA a potential asymmetric means to account for military gaps. For China, DI efforts appear central to military strategy and national efforts to project power and influence. For the United States, DI continues to serve as an enabler for military power.

Foundational assumptions for U.S. power projection may be shifting because of trends in DI. These assumptions include having a credible forward presence, rapidly flowing and sustaining forces, dominating domains, and minimal attrition and risk of escalation. Within the United States, DoD has historically developed networks and C3 capabilities associated with forward presence separate from DI, but trends suggest this may not be possible in the future.

Shaping the Long-Term U.S.-China Strategic Competition Requires a Comprehensive Understanding of Digital Infrastructure and Digital Presence

Military net assessments will benefit from an assessment of DI to account for the role of digital presence in military competition and conflict. Assessments to understand key characteristics of future warfare should also account for digital presence. Characterizing DoD's reliance on and relationship to DI and foundational elements is essential for informing an effective DoD strategy for long-term military competition.

Abbreviations

3GPP	3rd Generation Partnership Project
4G	fourth-generation
5G	fifth-generation
ADC	Asia Direct Cable
AI	artificial intelligence
AT&T	American Telephone and Telegraph
AU	African Union
B3W	Build Back Better World
BGP	Border Gateway Protocol
BRI	Belt and Road Initiative
C3	command, control, and communication
CCP	Chinese Communist Party
CEO	chief executive officer
COMSAT	communication satellite
DFC	International Development Finance Corporation
DI	digital infrastructure
DoD	U.S. Department of Defense
DSR	Digital Silk Road
EW	electronic warfare
FAANG	Facebook, Apple, Amazon, Netflix, and Google
FCC	Federal Communications Commission
IBM	International Business Machines
ICANN	Internet Corporation for Assigned Names and Numbers
IT	information technology
JADC2	Joint All-Domain Command and Control
MCI	Microwave Communications Incorporated
MOU	memorandum of understanding
NASA	National Aeronautics and Space Administration
NDI	national digital infrastructure

NSF	National Science Foundation
OAC	ownership of, access to, and control over
PC	personal computer
PLA	People's Liberation Army
PLAN	People's Liberation Army Navy
PoP	point of presence
PRC	People's Republic of China
R&D	research and development
RMA	revolution in military affairs
SEP	system essential patent
SMS	Short Message Service
TSMC	Taiwan Semiconductor Manufacturing Company
UK	United Kingdom
WTO	World Trade Organization
WWII	World War II
ZTE	Zhongxing Telecommunications Equipment Corporation

Bibliography

Aglionby, John, Emily Feng, and Yuan Yang, "African Union Accuses China of Hacking Headquarters," *Financial Times*, January 29, 2018.

Alberts, David S., John J. Garstka, and Frederick P. Stein, *Network Centric Warfare: Developing and Leveraging Information Superiority*, 2nd ed., Washington, D.C.: U.S. Department of Defense, 1999.

"Alcatel to Acquire Lucent in $13.4 Billion Deal," MSNBC, April 3, 2006.

"AT&T's History of Invention and Breakups," *New York Times*, February 13, 2016.

Augier, Mie, "Thinking About War and Peace: Andrew Marshall and the Early Development of the Intellectual Foundations for Net Assessment," *Comparative Strategy*, Vol. 32, No. 1, 2013, pp. 1–17.

Baldwin, David A., "Power and International Relations," in Walter Carlsnaes, Thomas Risse, and Beth A. Simmons eds., *Handbook of International Relations*, Newbury Park, Calif.: SAGE Publications, 2013, pp. 273–297.

Barrett, Jonathan, and Yew Lun Tian, "Pacific Undersea Cable Project Sinks After U.S. Warns Against Chinese Bid," Reuters, June 17, 2021.

Bechis, Francesco, "Undersea Cables: The Great Data Race Beneath the Oceans," webpage, Italian Institute for International Political Studies (ISPI), May 31, 2021.

Bibe, Miguel, "China Takes Over World Leadership in International Patent Applications," webpage, Inventa International, April 22, 2020.

Blair, Dennis C., written testimony, in U.S.-China Economic and Security Review Commission, "China's Military Power Projection and U.S. National Interests," hearing, Washington, D.C., February 20, 2020.

Bonds, Timothy M., James Bonomo, Daniel Gonzales, C. Richard Neu, Samuel Absher, Edward Parker, Spencer Pfeifer, Jennifer Brookes, Julia Brackup, Jordan Willcox, David R. Frelinger, and Anita Szafran, *America's 5G Era: Gaining Competitive Advantages While Securing the Country and Its People*, Santa Monica, Calif.: RAND Corporation, PE-A435-1, 2021. As of December 6, 2021:
https://www.rand.org/pubs/perspectives/PEA435-1.html

Bracken, Paul, "Net Assessment: A Practical Guide," *Parameters*, Vol. 36, No. 1, Spring 2006, pp. 90–100.

Brake, Doug, "Submarine Cables: Critical Infrastructure for Global Communications," Washington, D.C.: Information Technology & Innovation Foundation, April 2019.

Brands, Hal, *American Grand Strategy and the Liberal Order: Continuity, Change, and Options for the Future*, Santa Monica, Calif.: RAND Corporation, PE-209-OSD, 2016. As of December 6, 2021:
https://www.rand.org/pubs/perspectives/PE209.html

Burke, Edmund J., Kristen Gunness, Cortez A. Cooper III, and Mark Cozad, *People's Liberation Army Operational Concepts*, Santa Monica, Calif.: RAND Corporation, RR-A394-1, 2020. As of December 6, 2021:
https://www.rand.org/pubs/research_reports/RRA394-1.html

Cain, Geoffrey, *Samsung Rising: The Inside Story of the South Korean Giant That Set Out to Beat Apple and Conquer Tech*, New York: Penguin Random House, 2020.

Carter, Lionel, Douglas Burnett, Stephen Drew, Graham Marle, Lonnie Hagadorn, Deborah Bartlett-McNeil, and Nigel Irvine, *Submarine Cables and the Oceans: Connecting the World*, Cambridge, UK: United Nations Environment Programme, World Conservation Monitoring Centre, 2009. As of April 27, 2022:
http://www.iscpc.org/publications/icpc-unep_report.pdf

"China Telecom History," Funding Universe website, Company Profiles, undated. As of June 1, 2022:
http://www.fundinguniverse.com/company-histories/china-telecom-history/

China Telecom (Europe) Ltd., "History of China Telecom Europe," webpage, 2020.

Cliff, Roger, Mark Burles, Michael S. Chase, Derek Eaton, and Kevin L. Pollpeter, *Entering the Dragon's Lair: Chinese Antiaccess Strategies and Their Implications for the United States*, Santa Monica, Calif.: RAND Corporation, MG-524-AF, 2007. December 6, 2021:
https://www.rand.org/pubs/monographs/MG524.html

"Commercial Constellations Don't Live Up to the Hype: Euroconsult," *Satellite Pro Middle East*, December 13, 2020.

The Communications Act of 1934, as codified in 47 U.S.C. § 151 et seq.

Corera, Gordon, "How Britain Pioneered Cable-Cutting in World War One," BBC News, December 15, 2017.

Costello, John, and Joe McReynolds, *China's Strategic Support Force: A Force for a New Era*, Washington, D.C.: National Defense University Press, 2018.

Cozad, Mark R., and Nathan Beauchamp-Mustafaga, *People's Liberation Army Air Force Operations Over Water: Maintaining Relevance in China's Changing Security Environment*, Santa Monica, Calif.: RAND Corporation, RR-2057-AF, 2017. As of December 3, 2021:
https://www.rand.org/pubs/research_reports/RR2057.html

Cronin, Patrick M., Mira Rapp-Hooper, Harry Kresja, Alex Sullivan, and Rush Doshi, *Beyond the San Hai: The Challenge of China's Blue-Water Navy*, Washington, D.C.: Center for a New American Security, 2017.

Darby, Christopher, and Sarah Sewall, "The Innovation Wars: America's Eroding Technological Advantage," *Foreign Affairs*, March–April 2021.

David, Leonard, "China Lands on the Moon: Historic Robotic Lunar Landing Includes 1st Chinese Rover," Space.com, December 14, 2013.

DoD—*See* U.S. Department of Defense.

Dumat-ol Daleno, Gaynor, "CNMI Declares Emergency," *Pacific Daily News*, July 16, 2015.

Dutton, Peter A., Isaac B. Kardon, and Connor M. Kennedy, *Djibouti: China's First Overseas Strategic Strongpoint*, Newport, R.I.: U.S. Naval War College, April 2020.

Erwin, Sandra, "Defense Inspector General Finds Key Air Force Space Programs Vulnerable to Cyber Attacks, Sabotage," webpage, SpaceNews, August 16, 2018.

Eversden, Andrew, "Pentagon, Intel Partner to Make More US Microchips for Military," webpage, C4ISRnet, March 19, 2021.

Executive Research Associates Ltd., "China's Telecommunications Footprint in Africa," in Executive Research Associates Ltd., *China in Africa: A Strategic Overview*, Craighall, South Africa, October 2009.

FCC—*See* Federal Communications Commission.

Federal Communications Commission (FCC), "Foreign Ownership Rules and Policies," webpage, undated.

FitzGerald, Drew, "White House to Retool Pentagon Airwaves for 5G Networks," *Wall Street Journal*, August 10, 2020.

"Five Tech Giants Just Keep Growing," *Wall Street Journal*, May 1, 2021.

Fouquet, Helene, "China's 7,500-Mile Undersea Cable to Europe Fuels Internet Feud," Bloomberg, March 5, 2021.

Freedberg, Sydney J., "Making War a Software Problem: JAIC Director on JADC2," Breaking Defense website, September 8, 2020a.

———, "5G Experiments in US Pave Way to Battlefields Abroad," Breaking Defense website, November 4, 2020b.

Freifeld, Karen, "Biden Administration Adds New Limits on Huawei's Suppliers," Reuters, March 11, 2021.

Garafola, Cristina L., and Timothy R. Heath, *The Chinese Air Force's First Steps Toward Becoming an Expeditionary Air Force*, Santa Monica, Calif.: RAND Corporation, RR-2056-AF, 2017. As of December 6, 2021: https://www.rand.org/pubs/research_reports/RR2056.html

Garamone, Jim, "Joint All-Domain Command, Control Framework Belongs to Warfighters," press release, November 30, 2020.

Garamone, Jim, and Lisa Ferdinando, "DoD Initiates Process to Elevate U.S. Cyber Command to Unified Combatant Command," press release, Washington, D.C.: U.S. Department of Defense, August 18, 2017.

Goldman Sachs, "China Telecom Privatization Shines Through the Shadow of the Asian Financial Crisis," webpage, undated. As of April 27, 2022: https://www.goldmansachs.com/our-firm/history/moments/1997-china-telecom-privatization.html

Gore, Al, remarks, Inauguration of the First World Telecommunication Development Conference (WTDC-94), International Telecommunication Union, Buenos Aires, March 21, 1994.

Green, Michael J., *By More Than Providence: Grand Strategy and American Power in the Asia Pacific Since 1783*, New York: Columbia University Press, 2017.

Griffith, Keith, "Nokia to Acquire Alcatel-Lucent for $16.6 Billion," webpage, SDxCentral, April 15, 2015.

Grissom, Adam, "The Future of Military Innovation Studies," *Journal of Strategic Studies*, Vol. 29, No. 5, October 2006, pp. 905–934.

Gruss, Mike, "Pentagon: Narrowband? Wideband? Just Call Them Communications Satellites," webpage, SpaceNews, March 8, 2016.

Gunness, Kristen, written testimony, in U.S.-China Economic and Security Review Commission, "China's Military Power Projection and U.S. National Interests," hearing, Washington, D.C., February 20, 2020.

Gurman, Mark, Debby Wu, and Ian King, "Apple Aims to Sell Macs with Its Own Chips Starting in 2021," Bloomberg, April 23, 2020.

Harold, Scott W., and Justin Hodiak, "China's Semiconductor Industry: Autonomy Through Design?" Institut Montaigne website, September 25, 2020. As of December 5, 2021: https://www.institutmontaigne.org/en/blog/chinas-semiconductor-industry-autonomy-through-design

Harold, Scott W., and Rika Kamijima-Tsunoda, "Winning the 5G Race with China: A U.S.-Japan Strategy to Trip the Competition, Run Faster, and Put the Fix In," *Asia Policy*, Vol. 16, No. 3, July 2021.

Harwit, Eric, "Building China's Telecommunications Network: Industrial Policy and the Role of Chinese State-Owned, Foreign and Private Domestic Enterprises," *China Quarterly*, Vol. 190, June 2007, pp. 311–332.

Heath, Timothy R., *China's Pursuit of Overseas Security*, Santa Monica, Calif.: RAND Corporation, RR-2271-OSD, 2018. As of December 6, 2021: https://www.rand.org/pubs/research_reports/RR2271.html

Heath, Timothy R., Derek Grossman, and Asha Clark, *China's Quest for Global Primacy: An Analysis of Chinese International and Defense Strategies to Outcompete the United States*, Santa Monica, Calif.: RAND Corporation, RR-A447-1, 2021. As of December 6, 2021: https://www.rand.org/pubs/research_reports/RRA447-1.html

Hecht, Jeff, "Submarine Cable Goes for Record: 144,000 Gigabits from Hong Kong to L.A. in 1 Second," *IEEE Spectrum*, January 3, 2018.

Hicks, Kathleen, "Creating Data Advantage," memorandum for Senior Pentagon Leadership, Commanders of the Combatant Commands, Defense Agency and DoD Field Activity Directors, Washington, D.C.: U.S. Department of Defense, May 5, 2021.

Hitchens, Theresa, "Air Force to Launch 4G LTE at 20 More Bases Next Month," Breaking Defense website, November 12, 2020.

Hodiak, Justin, and Scott W. Harold, "Can China Become the World Leader in Semiconductors?" *The Diplomat*, September 25, 2020. As of December 5, 2021: https://thediplomat.com/2020/09/can-china-become-the-world-leader-in-semiconductors/

Hoehn, John R., "Joint All-Domain Command and Control (JADC2)," Washington, D.C.: Congressional Research Service, IF11493, June 4, 2021.

Huiss, Randy, *Proliferation of Precision Strike: Issues for Congress*, Washington, D.C.: Congressional Research Service, R42539, May 14, 2012.

Intel, "Intel at 50: The 8086 and Operation Crush," press release, May 31, 2018.

Johnson, Keith, and Elias Groll, "The Improbable Rise of Huawei," *Foreign Policy*, April 3, 2019.

Kostecka, Daniel J., "Places and Bases: The Chinese Navy's Emerging Support Network in the Indian Ocean," *Naval War College Review*, Vol. 64, No. 1, Winter 2011.

Krepinevich, Andrew F., *The Military-Technical Revolution: A Preliminary Assessment*, Washington, D.C.: Office of Net Assessment, July 1992.

Lenovo, "Lenovo Marks Decade of Success Since Acquisition of IBM's PC Business," press release, April 30, 2015.

Lewis, James Andrew, "5G: The Impact on National Security, Intellectual Property, and Competition," statement before the Senate Committee on the Judiciary, Washington, D.C., May 14, 2019.

Macias, Amanda, "China Is Quietly Conducting Electronic Warfare Tests in the South China Sea," CNBC, July 5, 2018.

Mackinnon, Amy, "For Africa, Chinese-Built Internet Is Better Than No Internet at All," *Foreign Policy*, March 19, 2019.

Markowitz, Jonathan N., and Christopher J. Farriss, "Power, Proximity, and Democracy: Geopolitical Competition in the International System," *Journal of Peace Research*, Vol. 55, No. 1, 2018.

Mazarr, Michael J., *Summary of the Building a Sustainable International Order Project*, Santa Monica, Calif.: RAND Corporation, RR-2397-OSD, 2018. As of December 6, 2021:
https://www.rand.org/pubs/research_reports/RR2397.html

Mazarr, Michael J., Jonathan Blake, Abigail Casey, Tim McDonald, Stephanie Pezard, and Michael Spirtas, *Understanding the Emerging Era of International Competition: Theoretical and Historical Perspectives*, Santa Monica, Calif.: RAND Corporation, RR-2726-AF, 2018. As of December 6, 2021:
https://www.rand.org/pubs/research_reports/RR2726.html

Mazarr, Michael J., Bryan Frederick, John J. Drennan, Emily Ellinger, Kelly Eusebi, Bryan Rooney, Andrew Stravers, and Emily Yoder, *Understanding Influence in the Strategic Competition with China*, Santa Monica, Calif.: RAND Corporation, RR-A290-1, 2021. As of December 6, 2021:
https://www.rand.org/pubs/research_reports/RRA290-1.html

Mazarr, Michael J., Timothy R. Heath, and Astrid Stuth Cevallos, *China and the International Order*, Santa Monica, Calif.: RAND Corporation, RR-2423-OSD, 2018. As of December 6, 2021:
https://www.rand.org/pubs/research_reports/RR2423.html

McCauley, Kevin, written testimony, in U.S.-China Economic and Security Review Commission, "China's Military Power Projection and U.S. National Interests," hearing, Washington, D.C., February 20, 2020.

Melgar, Luis, and Ana Rivas, "Biden's Infrastructure Plan Visualized: How the $2.3 Trillion Would Be Allocated," *Wall Street Journal*, April 1, 2021.

Meservey, Joshua, "Government Buildings in Africa Are a Likely Vector for Chinese Spying," backgrounder, Washington, D.C.: Heritage Foundation, May 20, 2020.

Mooney, John, "Russian Submarines 'Target Subsea Cables' off Coast of Kerry," *The Times* (London), August 16, 2020.

Motorola, *A Timeline Overview of Motorola History: 1928–2008*, 2008. As of December 6, 2021:
https://www.landley.net/history/mirror/6800_MotDoc.pdf

Moylan, Tom, "Trans-Atlantic Telephone Cable Will Contain a Piece of L.V.," The Morning Call website, September 20, 1987.

Murray, Williamson R., and Allan R. Millett, eds., *Military Innovation in the Interwar Period*, Cambridge, UK: Cambridge University Press, 1996.

Nakashima, Ellen, and Gerry Shih, "China Builds Advanced Weapons Systems Using American Chip Technology," *Washington Post*, April 7, 2021.

Nantulya, Paul, written testimony, in U.S.-China Economic and Security Review Commission, "China's Military Power Projection and U.S. National Interests," hearing, Washington, D.C., February 20, 2020.

Naraine, Ryan, "Nortel Hacking Attack Went Unnoticed for Almost 10 Years," ZDNet, February 14, 2012.

NASA—*See* National Aeronautics and Space Administration.

National Aeronautics and Space Administration, "Chang'e 1," webpage, NASA Science: Solar System Exploration, April 5, 2019a. As of June 5, 2021:
https://solarsystem.nasa.gov/missions/change-1/in-depth

———, "Viking 1," webpage, NASA Science: Solar System Exploration, July 16, 2019b. As of December 6, 2021:
https://solarsystem.nasa.gov/missions/viking-1/in-depth/

———, "Mars Pathfinder," webpage, NASA Science: Solar System Exploration, July 25, 2019c. As of December 6, 2021:
https://solarsystem.nasa.gov/missions/mars-pathfinder/in-depth/

———, "Mariner 9," webpage, NASA Science: Solar System Exploration, July 31, 2019d. As of December 6, 2021:
https://solarsystem.nasa.gov/missions/mariner-09/in-depth/

———, "Surveyor 1," webpage, NASA Science: Solar System Exploration, August 11, 2019e. As of December 6, 2021:
https://solarsystem.nasa.gov/missions/surveyor-1/in-depth/

———, "Lunar Orbiter 1," webpage, NASA Science: Solar System Exploration, September 12, 2019f. As of July 15, 2021:
https://solarsystem.nasa.gov/missions/lunar-orbiter-1/in-depth/

———, Office of the Inspector General, Office of Audits, *Cybersecurity Management and Oversight at the Jet Propulsion Laboratory*, Washington, D.C., IG-19-022, June 18, 2019. As of July 13, 2021:
https://oig.nasa.gov/docs/IG-19-022.pdf

National Air and Space Museum, "Discoverer/Corona: First U.S. Reconnaissance Satellite," SpaceRace website, 2002. As of July 15, 2021: https://airandspace.si.edu/exhibitions/space-race/online/sec400/sec420.htm

National Research Council, *Ada and Beyond: Software Policies for the Department of Defense*, Washington, D.C.: National Academy Press, 1997.

National Science Foundation, "A Brief History of NSF and the Internet," fact sheet, August 13, 2003. As of August 20, 2021: https://www.nsf.gov/news/news_summ.jsp?cntn_id=103050

Nokia Bell Labs, "Telstar 1," webpage, undated. As of July 13, 2021: https://www.bell-labs.com/about/history/innovation-stories/telstar-1/

Nouwens, Meia, *China's Digital Silk Road: Integration into National IT Infrastructure and Wider Implications for Western Defence Industries*, London: International Institute for Strategic Studies, February 2021.

NSF—*See* National Science Foundation.

Ochmanek, David, *Restoring U.S. Power Projection Capabilities: Responding to the 2018 National Defense Strategy*, Santa Monica, Calif.: RAND Corporation, PE-260-AF, 2018. As of December 6, 2021: https://www.rand.org/pubs/perspectives/PE260.html

Office of the Secretary of Defense, *Military and Security Developments Involving the People's Republic of China 2020: Annual Report to Congress*, Washington, D.C.: U.S. Department of Defense, 2020.

Organisation for Economic Co-operation and Development, *China's Belt and Road Initiative in the Global Trade, Investment and Finance Landscape*, Paris, 2018.

Page, Jeremy, Gordon Lubold, and Rob Taylor, "Deal for Naval Outpost in Cambodia Furthers China's Quest for Military Network," *Wall Street Journal*, July 22, 2019.

Palumbo, Paul, "Undersea Fiber Network to Link China and U.S.," *Lightwave*, February 1, 1998.

Pearson, Natalie Obiko, "Did a Chinese Hack Kill Canada's Greatest Tech Company?" *Bloomberg Businessweek*, July 1, 2020. As of July 13, 2021: https://www.bloomberg.com/news/features/2020-07-01/did-china-steal-canada-s-edge-in-5g-from-nortel

Peltier, Chad, written testimony, in U.S.-China Economic and Security Review Commission, "China's Military Power Projection and U.S. National Interests," hearing, Washington, D.C., February 20, 2020.

Peltier, Chad, Tate Nurkin, and Sean O'Connor, *China's Logistics Capabilities for Expeditionary Operations*, Jane's for the U.S.–China Economic and Security Review Commission, 2020. As of June 1, 2022:
https://www.uscc.gov/sites/default/files/2020-04/China%20Expeditionary%20Logistics%20Capabilities%20Report.pdf

Peters, Heidi M., "Defense Primer: U.S. Defense Industrial Base," Washington, D.C.: Congressional Research Service, IF10548, January 22, 2021.

Planetary Society, "Tianwen-1 and Zhurong, China's Mars Orbiter and Rover," webpage, undated.

Poling, Gregory B., written testimony, in U.S.-China Economic and Security Review Commission, "China's Military Power Projection and U.S. National Interests," hearing, Washington, D.C., February 20, 2020.

Pollpeter, Kevin L., Michael S. Chase, and Eric Heginbotham, *The Creation of the PLA Strategic Support Force and Its Implications for Chinese Military Space Operations*, Santa Monica, Calif.: RAND Corporation, RR-2058-AF, 2017. As of December 6, 2021:
https://www.rand.org/pubs/research_reports/RR2058.html

Public Law 99-433, Goldwater-Nichols Department of Defense Reorganization Act of 1986, October 1, 1986.

Public Law 116-92, National Defense Authorization Act for Fiscal Year 2020, Subtitle D, United States Space Force Act, December 20, 2019.

Qiu, Winston, "Why It Is China's Turn to Lead the Submarine Cable Industry," Telecom Ramblings website, February 11, 2014.

Ratner, Ely, Daniel Kliman, Susanna V. Blume, Rush Doshi, Chris Dougherty, Richard Fontaine, Peter Harrell, Martijn Rasser, Elizabeth Rosenberg, Eric Sayers, Daleep Singh, Paul Scharre, Loren DeJonge Schulman, Neil Bhatiya, Ashley Feng, Joshua Fitt, Megan Lamberth, Kristine Lee, and Ainikki Riikonen, *Rising to the China Challenge: Renewing American Competitiveness in the Indo-Pacific*, Washington, D.C.: Center for a New American Security, December 2019.

Roblin, Sebastien, "Russian Submarines Could Be Tampering with Undersea Cables That Make the Internet Work," *National Interest*, September 27, 2020.

Royal Irish Academy, "Evolution of the Microchip," webpage, July 27, 2020. As of July 13, 2021:
https://www.ria.ie/news/library-library-blog/evolution-microchip

Saunders, Phillip C., *Beyond Borders: PLA Command and Control of Overseas Operations*, Washington, D.C.: National Defense University, July 2020.

Sbragia, Chad, written testimony, in U.S.-China Economic and Security Review Commission, "China's Military Power Projection and U.S. National Interests," hearing, Washington, D.C., February 20, 2020.

Scobell, Andrew, Edmund J. Burke, Cortez A. Cooper III, Sale Lilly, Chad J. R. Ohlandt, Eric Warner, and J. D. Williams, *China's Grand Strategy: Trends, Trajectories, and Long-Term Competition*, Santa Monica, Calif.: RAND Corporation, RR-2798-A, 2020. As of December 6, 2021: https://www.rand.org/pubs/research_reports/RR2798.html

Semiconductor Manufacturing International Corporation, "Overview," webpage, undated. As of August 18, 2021: https://www.smics.com/en/site/company_info

Sherman, Jason, "Microelectronics Is DOD's New No. 1 Technology Priority, Bumping Hypersonics to No. 3," *Inside Defense*, June 29, 2020.

Singer, Peter W., "The Lessons of World War 3," statement prepared for the U.S. Senate Committee on Armed Services, hearing on "Future of Warfare," 114th Cong., 1st Sess., Senate Hearing 114-211, Washington, D.C., November 3, 2015.

Spegele, Brian, and Kate O'Keeffe, "China Maneuvers to Snag Top-Secret Boeing Satellite Technology," *Wall Street Journal*, December 4, 2018.

———, "China Exploits Fleet of U.S. Satellites to Strengthen Police and Military Power," *Wall Street Journal*, April 23, 2019.

Steel in the Air, Inc., "The Wireless Carrier Timeline & Industry Evolution: Pre-1983–Present," webpage, 2014.

Stone, Alex, and Peter Wood, *China's Military—Civil Fusion Strategy: A View from Chinese Strategists*, Montgomery, Ala.: China Aerospace Studies Institute, 2020.

Sutter, Karen M., "'Made in China 2025' Industrial Policies: Issues for Congress," Washington, D.C.: Congressional Research Service, IF10964, August 11, 2020.

Tanenbaum, Andrew S., *Computer Networks*, 2nd ed., Hoboken, N.J.: Prentice Hall, 1989.

TeleGeography, "Submarine Cable Map," webpage, last updated November 15, 2021.

"Top 10 Satellite Manufacturers in the Global Space Industry 2018," *Technavio Blog*, October 9, 2018.

"Trans-Pacific Submarine Cable Systems," webpage, Submarine Cable Networks, undated.

Triolo, Paul, Kevin Allison, Clarise Brown, and Kelsey Broderick, "The Digital Silk Road: Expanding China's Digital Footprint," Washington, D.C.: Eurasia Group, April 8, 2020.

TSMC, "Company Info," webpage, undated. As of April 27, 2022: https://www.tsmc.com/english/aboutTSMC/company_profile

U.S.-China Economic and Security Review Commission, "China's Military Power Projection and U.S. National Interests," hearing, Washington, D.C., February 20, 2020. As of June 1, 2022:
https://www.uscc.gov/hearings/
chinas-military-power-projection-and-us-national-interests

U.S. Department of Defense, *Summary of the 2018 National Defense Strategy of the United States of America: Sharpening the American Military's Competitive Edge*, Washington, D.C., 2018.

———, *Department of Defense 5G Strategy Implementation Plan: Advancing 5G Technology & Applications Securing 5G Capabilities*, Washington, D.C., December 17, 2020.

U.S. Government Accountability Office, *DOD Management Approach and Processes Not Well-Suited to Support Development of Global Information Grid*, Washington, D.C., GAO-06-211, January 2006.

U.S. Senate, Permanent Subcommittee on Investigations, *Threats to U.S. Networks: Oversight of Chinese Government-Owned Carriers*, staff report, Washington, D.C., June 9, 2020. As of December 6, 2021:
https://www.hsgac.senate.gov/imo/media/doc/2020-06-09%20PSI%20
Staff%20Report%20-%20Threats%20to%20U.S.%20Communications%20
Networks.pdf

Vergun, David, "DoD Looking for Advanced Command, Control Solution," press release, Washington, D.C.: U.S. Department of Defense, June 4, 2021.

Vodafone, "Where We Operate," webpage, undated.

Wall, Mike, "China's First Mars Mission, Tianwen-1, Successfully Enters Orbit Around Red Planet," Space.com, February 10, 2021.

Waltz, Kenneth N., *Theory of International Politics*, reprint, Long Grove, Ill.: Waveland Press, Inc., 2010.

Watson, Cynthia, written testimony, in U.S.-China Economic and Security Review Commission, "China's Military Power Projection and U.S. National Interests," hearing, Washington, D.C., February 20, 2020.

Webb, Alex, "A Few Glass Strands Will Protect U.S. Tech From China," Bloomberg, April 5, 2021.

Weinbaum, Cortney, "The Intelligence Community's Deadly Bias Toward Classified Sources," Defense One website, April 9, 2021.

Weinbaum, Cortney, Steven Berner, and Bruce McClintock, *SIGINT for Anyone: The Growing Availability of Signals Intelligence in the Public Domain*, Santa Monica, Calif.: RAND Corporation, PE-273-OSD, 2017. As of December 6, 2021:
https://www.rand.org/pubs/perspectives/PE273.html

Weitering, Hanneke, "China's 1st Mars Rover 'Zhurong' Lands on the Red Planet," Space.com, May 14, 2021. As of July 15, 2021:
https://www.space.com/china-mars-rover-landing-success-tianwen-1-zhurong

Wen, Yun, *The Huawei Model: The Rise of China's Technology Giant*, Chicago: University of Illinois Press, 2020.

White House, *National Security Strategy of the United States of America*, Washington, D.C., December 2017.

———, "President Biden's Bipartisan Infrastructure Law," press release, November 2021. As of April 27, 2022:
https://www.whitehouse.gov/bipartisan-infrastructure-law/

Williams, Brad D., "FCC's Carr: Close Chinese Backdoors into U.S. Networks," Breaking Defense website, April 1, 2021.

Williams, Ian, and Masao Dahlgren, "More Than Missiles: China Previews Its New Way of War," Washington, D.C.: Center for Strategic and International Studies, October 2019.

Winkler, Jonathan Reed, "Silencing the Enemy: Cable-Cutting in the Spanish-American War," War on the Rocks website, November 6, 2015.

Woo, Stu, and Daniel Michaels, "China Buys Friends with Ports and Roads. Now the U.S. Is Trying to Compete," *Wall Street Journal*, July 15, 2021.

World Bank Group, *Belt and Road Economics: Opportunities and Risks of Transport Corridors*, Washington, D.C., 2019.

Wortham, Jenna, "Nokia Siemens Agrees to Pay Cash for Division of Motorola," *New York Times*, July 19, 2010.

Xiao, Eva, "China Passes One of the World's Strictest Data-Privacy Laws," *Wall Street Journal*, August 20, 2021.

Yap, Chuin-Wei, "State Support Helped Fuel Huawei's Global Rise," *Wall Street Journal*, December 25, 2019.

Young Professionals in Foreign Policy, *US-China Futures: Briefing Book*, New York: Schmidt Futures, March 2021.

Yung, Christopher D., Ross Rustici, Scott Devary, and Jenny Lin, *"Not an Idea We Have to Shun": Chinese Overseas Basing Requirements in the 21st Century*, Washington, D.C.: National Defense University, 2014.